Harvesting Rainwater

A Sustainable Guide to Collecting, Storing, and Utilizing Nature's Gift for Water Conservation and Self-Sufficiency

Table of Contents

Introduction

In a world where water is an increasingly precious commodity, imagine a solution that conserves this vital resource and transforms a mundane rain shower into a sustainable lifeline for your home and garden. Welcome to the captivating realm of "Harvesting Rainwater."

A Watery Adventure

There'll come a day when you're no longer at the mercy of your water bill, your garden thrives without guzzling gallons from the faucet, and you're not left high and dry during droughts. This book is your ticket to a watery wonderland where rain becomes a partner in your journey to sustainability.

Purpose Unveiled

The primary goal of "Harvesting Rainwater" is to empower you with the knowledge and skills to harness the incredible potential of rainwater. More than a resource, rainwater is a solution, and this book is your roadmap to unlock its full potential. It's not just about water conservation but a lifestyle shift towards self-sufficiency, eco-friendliness, and a deeper connection with the environment.

Why This Book Stands Out

What sets this guide apart from the rest in the market? Simply put, it's designed with you in mind. There's no complicated jargon or intricate diagrams that make your head spin. This book is your friendly neighbor, inviting you over for a chat about rain, barrels, and sustainable living.

- **Easy to Understand:** Forget the perplexing technicalities. This book breaks down the rainwater harvesting process into bite-sized, easily digestible pieces. You don't need an engineering degree to understand the concepts presented here.

- **Great for Beginners:** Whether you're a gardening enthusiast or someone just dipping their toes into the sustainability pool, this book is the perfect starting point. It assumes no prior knowledge, guiding you from the basics to becoming a rainwater harvesting maestro.

- **Hands-On Methods and Instructions:** This isn't a theoretical schoolbook. It's a hands-on manual. Dive into practical, step-by-step instructions that turn theory into action. By the time you put it down, you'll be ready to implement your rainwater harvesting system confidently.

- **Engaging and Accessible:** Are complicated manuals collecting dust on your shelf? This one won't join them. "Harvesting Rainwater" is a page-turner, written in a style that's engaging, humorous, and accessible.

Say goodbye to water woes and join the movement towards a more sustainable future. This book is a key to a greener, more self-sufficient tomorrow. So, dive into the pages of "Harvesting Rainwater" and let the water revolution begin. Your garden, your wallet, and the planet will thank you.

Chapter 1: The Basics of Rainwater Harvesting

Rainwater has played a vital role throughout human history. From ancient civilizations to the environmentally conscious present, rainwater harvesting has woven itself into the fabric of sustainable living. In this chapter, you'll go through the fundamental concepts of rainwater collection. You'll be tracing its roots in history, understanding its modern significance, and exploring its place within the intricate web of the natural water cycle.

Rainwater has played a vital role throughout human history.
https://www.pexels.com/photo/raindrops-1529360/

Historical Context of Rainwater Harvesting

Rain, the ageless dance of droplets from the heavens, has been an eternal companion to humanity. In the intricate choreography of nature, rainwater has been more than a fleeting visitor. It's a timeless resource that early civilizations, with their keen understanding of the environment, ingeniously harnessed for survival. Prepare to travel back in time to explore the historical context of rainwater harvesting and witness the evolution of techniques that have shaped this practice.

Unveiling the Ingenuity of the Nabataeans

The Nabataeans, inhabitants of the ancient city of Petra, were true maestros of rainwater harvesting. Situated in the heart of a desert, their survival depended on their ability to maximize every precious drop of rain. The ingenious rain gardens of Petra were a testament to their advanced understanding of water flow and conservation.

Carved into the rose-red sandstone, the Nabataean dams and cisterns formed an intricate network designed to capture and direct rainwater. These structures weren't just utilitarian. They were a marriage of functionality and artistry. The Nabataeans sculpted their environment to harmonize with the sporadic yet life-giving rains, showcasing a level of ingenuity that still captivates modern minds.

Their approach was proactive. They didn't wait for water scarcity to force innovation, anticipating the needs of their community and engineering solutions for sustained prosperity. The legacy of the Nabataeans serves as a reminder that, even in the harshest environments, humans have the potential to transform challenges into opportunities.

Greek Wisdom in Rooftop Catchment

The Greeks, renowned for their contributions to philosophy and science, also recognized the value of rainwater. In a society that esteemed wisdom, they implemented sophisticated rooftop catchment systems to capture and channel rain into storage vessels.

The Greeks understood that rain was a source of life. The rooftop catchment systems, often seen in conjunction with the architecture of ancient Greek homes, were a practical manifestation of their reverence for water. By collecting rainwater, the Greeks ensured a reliable supply for domestic and agricultural use.

This harmonious integration of practicality and philosophy reflected a holistic approach to living in concert with nature. The Greeks, while in their pursuit of knowledge, acknowledged the interconnectedness of human life with the environment. The wisdom of rooftop catchment systems wasn't just a technological achievement. It was a manifestation of a deeper understanding of the symbiotic relationship between humanity and the elements.

Mohenjo-Daro's Hydraulic Brilliance

The ancient city of Mohenjo-Daro, nestled in the fertile plains of the Indus Valley, showcases another chapter in the historical saga of rainwater harvesting. The inhabitants of this advanced civilization were pioneers in crafting intricate systems of canals and reservoirs to harness the monsoon rains.

Mohenjo-Daro's strategic water management wasn't merely about survival but about thriving. The canals and reservoirs were carefully planned and engineered systems that sustained agricultural pursuits. The brilliance of Mohenjo-Daro lay in the architectural layout of the city and the foresight to leverage the seasonal abundance of rain for long-term prosperity.

The city's advanced hydraulic systems were a testament to a civilization's organizational and engineering prowess that flourished in harmony with its environment. The practical lessons from Mohenjo-Daro transcend time, reminding you that sustainable water management is not a modern concept but an age-old wisdom rooted in shared human history.

Roman Mastery: Aqueducts and Cisterns

The Romans, synonymous with engineering marvels, elevated rainwater harvesting to an art form. The grandeur of their aqueducts and cisterns supplied water for domestic use and played a crucial role in sustaining the expansive Roman Empire.

With their impressive arches spanning across landscapes, aqueducts were engineering feats that transported water over vast distances. Cisterns, strategically placed within cities and estates, stored rainwater for times of need. The Romans recognized that rainwater was a strategic asset that could be managed on a grand scale.

The mastery of the Romans extended beyond conquests. It encompassed the thoughtful utilization of natural resources. Their aqueducts and cisterns were conduits of sustainability, ensuring a stable water supply for a burgeoning civilization. The legacy of Roman rainwater harvesting is a testament to the enduring impact of farsighted environmental stewardship.

Evolution into the Middle Ages

As the medieval period unfolded, monasteries became hubs of innovation in water management. Vast rooftop systems collected rainwater, serving agricultural and domestic needs. Monks, often the custodians of knowledge and wisdom, recognized the value of rainwater for sustenance and the spiritual and communal well-being of their societies.

In the monastic tradition, rainwater harvesting went from a practical necessity to a spiritual practice. Monasteries often featured intricate systems of gutters and downspouts that directed rainwater into storage facilities. The collected rainwater, considered pure and untainted, was used for various purposes, including brewing and medicinal preparations.

The monastic approach to rainwater harvesting reflects a profound understanding of the interconnectedness of physical and spiritual well-being. More than survival, it was about holistic living. The echoes of medieval rainwater harvesting resonate in the quiet courtyards of monasteries, where the timeless practice merged with a deeper appreciation for the sanctity of water.

Renaissance Refinement

The Renaissance period witnessed a refinement of rainwater harvesting systems. Elaborate designs adorned the estates of the wealthy, reflecting a practical approach to water conservation and an aesthetic integration of functionality and beauty. The grandeur of these systems mirrored the cultural and artistic aspirations of the time.

As the Renaissance unfolded, a renewed interest in classical knowledge and a celebration of human potential spurred advancements in various fields. In the realm of rainwater harvesting, this era saw the fusion of artistic sensibilities with practical utility. Rooftop structures became ornate, featuring intricate carvings and designs that transformed the functional elements into works of art.

The refinement of rainwater harvesting during the Renaissance was a cultural expression. The estates of the wealthy became showcases of both technological prowess and artistic ingenuity. The convergence of beauty and utility in rainwater harvesting systems mirrored the broader

Renaissance spirit. It was an era when human achievement and all its facets were celebrated.

Resurgence in the Modern Era

Fast forward to the present, and you'll find yourself grappling with the challenges of a rapidly changing climate. The ancestral wisdom, however, has not been forgotten. There is a resurgence of interest in these ancient practices as modern societies seek sustainable solutions to contemporary problems. The echoes of rain gardens, rooftop catchments, and aqueducts from centuries past resonate as humans explore ways to harmonize needs with the environment.

In an era marked by technological advancements and a growing awareness of environmental concerns, the principles of rainwater harvesting are experiencing a renaissance. Water scarcity, climate change, and increasing urbanization have prompted a revisit to age-old practices that stood the test of time.

The resurgence of interest isn't merely a nostalgic look back. It's a strategic response to contemporary issues. Rainwater harvesting, once a necessity born out of survival, is now a choice. It's an informed decision to adopt sustainable water practices. The ancient techniques that allowed civilizations to flourish in diverse environments are becoming guiding lights in the quest for resilient and water-conscious communities.

Reviving Ancient Wisdom

The historical context of rainwater harvesting is an excellent example of human ingenuity and environmental stewardship. The lessons from the Nabataeans, Greeks, Indus Valley Civilization, and Romans are not relics of a bygone era. They are beacons guiding you toward a more sustainable future.

In your quest to address water scarcity and environmental challenges, you can draw inspiration from the evolution of rainwater harvesting techniques. The same principles that allowed ancient civilizations to thrive in diverse landscapes inform the contemporary efforts to build resilient and water-conscious communities.

As humanity stands at the crossroads of history and progress, the resurgence of interest in rainwater harvesting represents more than a nod to tradition. It's a conscious choice to embrace the wisdom of the past in shaping a sustainable and water-secure future. The droplets that fell on ancient civilizations continue to echo through time. They invite you to harness the liquid gold from the skies for the well-being of this planet and generations to come.

The Modern Impetus

In the unfolding narrative of the 21st century, the world finds itself on the brink of a water crisis. As populations increase, urban landscapes expand, and the capricious effects of climate change manifest with traditional water sources straining under pressure. In this era of uncertainty, rainwater harvesting emerges as a beacon of hope. It's a sustainable solution offering a reliable alternative to conventional water supplies.

Dwindling Water Resources

Water scarcity looms over the horizon in an era marked by relentless urbanization and a growing global population. Traditional water sources, like rivers and aquifers, face unprecedented stress. The water demand has surged to unparalleled levels, driven by the needs of industries, agriculture, and growing urban settlements. As these traditional sources strain to meet the demand, this water crisis necessitates innovative and sustainable alternatives.

The Significance of Rainwater Harvesting:

- **Sustainability**: Rainwater harvesting provides a sustainable alternative to overexploited traditional water reservoirs.

- **Replenishable Source**: Rainwater is a replenishable source that eases the burden of depleting water resources.

- **Conservation of Groundwater**: By capturing rainwater, you contribute to conserving precious groundwater and surface water reserves.

Environmental Responsibility

The modern impetus for rainwater harvesting extends beyond a response to water scarcity. It aligns with the growing wave of environmental responsibility sweeping across individuals and communities. As awareness of ecological issues deepens, people seek tangible ways to reduce their ecological footprint. Rainwater harvesting emerges as a tangible and impactful solution, presenting an opportunity to conserve water resources while minimizing the environmental impact associated with traditional water extraction methods.

The Environmental Impact of Rainwater Harvesting:

- **Reduced Ecological Footprint:** Rainwater harvesting reduces reliance on traditional water sources, minimizing the environmental impact of water extraction.

- **Preservation of Natural Ecosystems:** Every drop collected preserves natural ecosystems, keeping rivers undisturbed, aquifers naturally recharged, and aquatic habitats in balance.

- **Conscious Contribution:** Choosing rainwater harvesting is a conscious contribution to the broader sustainability of the planet.

Self-Sufficiency

The desire for self-sufficiency acts as a powerful motivator, prompting many to explore the realm of rainwater harvesting. By capturing rainwater on their premises, individuals gain a degree of independence from municipal water supplies. This newfound autonomy offers a reliable water source and contributes to reducing the burden on centralized water distribution systems.

Self-Sufficiency through Rainwater Harvesting:

- **Autonomy from Municipal Supplies:** Rainwater harvesting provides individuals with a reliable water source, reducing reliance on municipal supplies.

- **Agricultural Independence:** Rainwater becomes a valuable asset in agriculture, fostering self-sufficiency in nurturing crops and sustaining livestock.

- **Community Resilience:** The ethos of self-sufficiency extends to entire communities, reducing reliance on external water sources and fostering a more sustainable and resilient way of life.

A Holistic Approach to Water Security

The modern impetus for rainwater harvesting is multifaceted, addressing immediate water scarcity concerns and embracing environmental responsibility in the pursuit of self-sufficiency. As you navigate the complexities of the 21st century, rainwater harvesting emerges as a technological solution and a holistic approach to water security. This approach harmonizes with the environment, preserves natural ecosystems, and empowers individuals and communities to take charge of their water future.

The Holistic Vision of Rainwater Harvesting:

- **Harmonizing with the Environment:** Rainwater harvesting harmonizes with the environment, preserving natural ecosystems and contributing to the sustainability and preservation of the planet.
- **Empowering Individuals and Communities:** By choosing rainwater harvesting, individuals and communities progress toward a more sustainable and water-secure world.
- **A Flourishing Tomorrow:** Rainwater becomes a source of empowerment, fostering a connection between humanity and nature, shaping a flourishing and water-wise tomorrow.

Rainwater in the Natural Water Cycle

To truly appreciate the art of rainwater harvesting, you must first immerse yourself in the poetic process of the natural water cycle. This intricate choreography unfolds with the sun's tender embrace, coaxing moisture from oceans, lakes, and rivers into the sky through the enchanting process of evaporation.

The Dance of Droplets

The journey of rainwater begins high above, where the sun's warmth becomes a catalyst for transformation. Oceans, lakes, and rivers surrender their liquid essence to the sky, rising as invisible water vapor. As water vapor ascends, it merges into clouds. This cloudy collaboration is a testament to nature's artistry, a prelude to the grand performance that awaits.

The clouds gather and disperse, carrying the promise of life-giving rain. The process continues as these clouds weave intricate patterns influenced by atmospheric currents and temperature variations. When conditions align, the clouds release their gathered moisture in a cascade of rain. Isn't it a sublime spectacle that sustains life on Earth?

Cloudy Collaboration

This collaboration reaches its crescendo as clouds release their watery payload in the form of droplets. Rainfall is a fundamental act in nature's recycling system. Raindrops descend to the earth, refreshing the land, replenishing rivers, and refilling aquifers.

The rain-saturated earth becomes a stage for life's renewal. Seeds sprout, rivers flow, and ecosystems flourish in response. The rain's journey, however, is far from over. Its impact resonates in a continuous cycle, sustaining life and maintaining the delicate equilibrium of this planet.

Raindrops on leaves and soil cleanse the atmosphere and wash away dust and pollutants. The earthy fragrance that arises when rain meets dry soil is a testament to this purifying dance. The very act of rain falling to the ground is nature's way of rejuvenating and purifying the environment.

The Resilience of Rainwater

Unlike water from traditional sources, rainwater carries a simplicity that makes it an appealing alternative for various purposes. Its innate softness renders it ideal for nurturing plants, while its lack of mineral content deems it preferable for certain domestic uses.

The journey from sky to earth imbues rainwater with a unique character. As it descends, it acts as nature's purifier, cleansing itself of impurities acquired during its voyage. This innate resilience and purity make rainwater a versatile resource. It's a liquid canvas awaiting human ingenuity to paint its purpose.

The very composition of rainwater, with its lack of mineral content and low levels of dissolved solids, distinguishes it from other water sources. This purity makes it suitable for irrigation and household use. Its purity also positions it as an ideal source for certain industrial applications.

Rainwater, being free from the impurities found in ground or surface water, reduces the need for complex filtration processes. This simplicity in composition enhances its usability, all while reducing the energy and resources required to make it usable for various purposes.

The Sustainability Quotient

Rainwater harvesting isn't just about human intervention. It's a harmonious integration into this natural symphony. Collecting rainwater allows you to actively participate in the cycle without disrupting its delicate balance. It's a sustainable choice that acknowledges the interconnectedness of all elements in Earth's ecological theater.

The sustainability quotient of rainwater harvesting lies in its practical applications and alignment with the rhythms of nature. It's a choice that transcends individual needs. It reflects the ancient wisdom of civilizations that understood the droplets long before modern aspirations.

Harmony with Nature

In the grand narrative of this planet, rainwater plays a crucial role as both performer and protagonist. Its journey, from vapor to cloud to raindrop, is a testament to the resilience and interconnectedness of Earth's systems. By embracing rainwater harvesting, you embrace a harmonious relationship with nature. This relationship goes beyond mere resource utilization to a profound understanding and stewardship of the intricate water cycle.

As you collect rainwater, you become the choreographer of a sustainable future. Each collected raindrop is a step toward preserving the environment's delicate balance. With its simplicity and resilience, rainwater invites you to join the orchestra of conscious living, where every action contributes to the well-being of the planet you call home.

Integrating rainwater harvesting into your daily life is a declaration that you are not separate from nature but an integral part of its rhythms. This practice aligns with the principles of permaculture. Its philosophy mimics natural ecosystems to create sustainable and regenerative habitats suitable for humans.

The simplicity of rainwater harvesting systems, often consisting of gutters, downspouts, and storage containers, mirrors the elegance of nature's processes. This simplicity, coupled with its profound impact on local water resources, reinforces the idea that sustainability is not about complex solutions but about working with the inherent gifts of the natural world.

In the following chapters, you'll dive deeper into the practical aspects of rainwater harvesting. From the tools needed to the step-by-step processes, you will become an active participant in the age-old tradition of collecting and utilizing nature's liquid gold. As you unveil the secrets of rainwater harvesting, you'll discover the power to shape a more sustainable and self-sufficient future, one raindrop at a time.

Chapter 2: The Science Behind Precipitation

From the gentle drizzle that nurtures the soil to the torrential downpour that shapes landscapes, the science behind precipitation is a fascinating journey through the heart of the water cycle. This chapter unravels the foundational meteorological principles governing rain formation and fall, exploring the intricate dance of water molecules as they traverse the atmosphere.

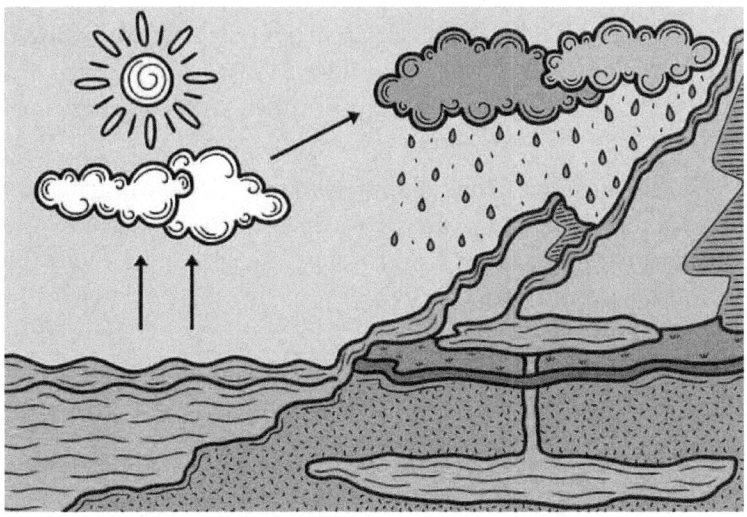

The water cycle, a mesmerizing process orchestrated by the forces of nature, is a perpetual dance that sustains life on this planet.
https://pixabay.com/zh/illustrations/water-cycle-rain-clouds-8176128/

The Water Cycle Demystified

The water cycle, a mesmerizing process orchestrated by the forces of nature, is a perpetual dance that sustains life on this planet. At its core are three captivating acts: evaporation, condensation, and precipitation. Here's a closer look at the intricacies of each stage to demystify the awe-inspiring journey of water molecules as they traverse the vast expanse of the atmosphere.

Evaporation

At the heart of the water cycle lies the enchanting process of evaporation. This act unfolds under the tender caress of the sun's warming rays. Water, in its liquid form, experiences a magical transformation into vapor. This metamorphosis is more than a scientific phenomenon. It is a fascinating dance of molecules. It's a poetic interplay between the liquid surface of oceans, lakes, and rivers and the beckoning call of the sun.

The Dance of Molecules

- **Solar Embrace:** The sun extends its golden fingers across the Earth's surface, imparting kinetic energy to water molecules. This solar embrace is where water molecules gain the energy to liberate themselves from liquid form.

- **Escape to the Skies:** With newfound energy, water molecules shed their liquid form and ascend into the atmosphere. This ethereal ascent marks the start of a journey that transcends geographical boundaries and embraces the boundless expanse above.

- **Global Voyage:** Once liberated, these water vapor molecules go on a global voyage, carried by air currents and wind. From the balmy tropics to the frigid poles, the liberated vapor becomes an intrepid traveler, ready to engage in the next act of the water cycle.

The Atmospheric Odyssey

- **Air Currents and Wind:** The liberated water vapor becomes a passenger on air currents and wind, creating a dynamic aerial phenomenon. These atmospheric currents carry water vapor across vast distances, shaping the atmospheric dynamics that influence weather patterns.

- **Moisture Reservoirs:** The vapor, now suspended in the atmosphere, forms moisture reservoirs that hold the potential for future precipitation. These reservoirs, invisible to the naked eye, are essential contributors to the delicate balance that sustains life on Earth.
- **Interconnected Systems:** The atmospheric odyssey of water vapor is part of a complex and interconnected system that influences climate, weather, and the distribution of water resources across the globe.

Condensation

As the vapor molecules ascend into the atmosphere, they encounter cooler air at higher altitudes. This encounter triggers a majestic transformation, the act of condensation. In it, vapor surrenders its ephemeral form, condensing into tiny droplets or ice crystals. The newly formed water particles then gather around particles like dust or aerosols, combining to create the canvas upon which clouds paint their ethereal beauty across the sky.

The Symphony of Condensation

- **Temperature:** The change in temperature at higher altitudes is the catalyst of this stage. Cooler air encourages the vapor molecules to slow down and embrace their liquid form once again.
- **Gathering in Clouds:** As condensation takes hold, these minuscule water particles dance around atmospheric particles, forming clouds. These clouds, in their myriad shapes and sizes, become the visual poetry of the sky, reflecting the essence of the atmosphere.
- **Aerial Artistry:** The resulting clouds, whether wispy cirrus or dense cumulus, capture and reflect the ever-changing moods of the atmosphere. This act of condensation decorates the sky and sets the stage for the grand finale of precipitation.

Cloud Formations and Aesthetic Revelry

- **Diverse Cloud Types:** Condensation gives rise to an array of cloud types, each with its unique characteristics. Cirrus clouds are high and wispy, while cumulonimbus clouds are towering and majestic, heralding the potential for intense precipitation.

- **Weather Indicators:** Cloud formations serve as invaluable indicators of imminent weather changes. Understanding the nuances of cloud aesthetics allows meteorologists and weather enthusiasts to decipher atmospheric conditions and predict upcoming precipitation events.
- **Artistic Splendor:** The aesthetic revelry of cloud formations is a testament to the creative artistry of nature. From radiant sunsets reflected in altocumulus clouds to the ominous beauty of an approaching storm in nimbostratus clouds, condensation transforms the sky into a canvas of ever-changing masterpieces.

Precipitation

The grand culmination of the water cycle's atmospheric journey is the act of precipitation. It occurs when the condensed water droplets within clouds grow heavy enough to overcome the resistance of air currents. Under the influence of gravity, they descend earthward, transforming into varied forms of precipitation, including rain, snow, sleet, or hail.

The Dramatic Descent

- **Growing Heavier:** Within the clouds, the water droplets continue to grow in size as they collide and merge. This growth transforms them into precipitation – ready to make its descent.
- **Gravity's Pull:** The moment arrives when these condensed droplets become too weighty for the air to support. Gravity, the omnipotent force, pulls them downward, initiating the descent that defines precipitation.
- **Elixir of Life:** As these droplets kiss the Earth, they contribute to the vital cycle of life. Whether nurturing the soil, replenishing lakes and rivers, or sustaining ecosystems, precipitation is the elixir that rejuvenates and sustains this planet.

The Holistic Impact

- **Soil Nourishment:** Precipitation seeps into the soil, providing essential hydration to plant roots. This nourishment is fundamental to the growth and vitality of terrestrial ecosystems.
- **Aquatic Replenishment:** Lakes, rivers, and oceans receive a replenishing embrace from precipitation. This influx of freshwater sustains aquatic habitats, maintaining the delicate balance of marine ecosystems.

- **Ecosystem Resilience:** The holistic impact of precipitation extends beyond individual components of the environment. It contributes to the resilience of ecosystems, ensuring the continued vitality and diversity of life on Earth.

The Harmonious Cycle

The journey from evaporation to precipitation is not just a linear progression. It's a harmonious cycle that perpetuates life on Earth. Each act in this atmospheric dance is interconnected, creating a seamless choreography that repeats itself endlessly. From the liquid embrace of evaporation to the artistic formations of condensation and the dramatic descent of precipitation, the water cycle is a living, breathing testament to the interconnectedness of nature.

The Enchanting Finale

As you demystify the water cycle, you unveil the intricate beauty that sustains life on this planet. This perpetual dance, conducted by the sun, the atmosphere, and the Earth, is a testament to the resilience and interconnectedness of nature. As you gaze upon the clouds, feel the rain on your face, and witness the cycle unfold around you, you are not a mere spectator but an active participant in the grand symphony that is the water cycle.

Factors Influencing Rainfall

In the intricate workings of rainfall, nature conducts a symphony where various factors harmonize to create the delicate dance of precipitation. Each element plays a crucial role, from temperature fluctuations and the topographical stage upon which rain unfolds to the nuanced air currents and humidity that set the rhythm. In this section, you'll learn about the complex interplay of factors influencing rainfall dynamics, deciphering the poetry written in raindrops.

Temperature Fluctuations

Temperature holds the baton that directs the rhythm of precipitation patterns. Its influence is profound, shaping the atmospheric conditions that give rise to rainfall. The balance between temperature fluctuations is a key to deciphering the dynamics of rain.

The Dance of Warm and Cool

- **Warmer Air, Increased Moisture:** In the atmosphere, warmer air leads to increased moisture retention. As temperatures rise, air gains the capacity to hold more water vapor through evaporation. It sets the stage for heightened evaporation from oceans, lakes, and other water bodies, fostering the birth of clouds.

- **Cooler Temperatures, Condensation Groundwork:** On the contrary, cooler temperatures provide the backdrop for condensation to take center stage. When warm, moisture-laden air encounters cooler conditions, it reaches its dew point (the temperature at which condensation occurs). This transition from vapor to liquid lays the groundwork for cloud formation and, eventually, precipitation.

Understanding the balance between these temperature fluctuations unveils the intricate dance of rainfall dynamics. From the initial evaporation to the eventual condensation and precipitation, temperature is the guiding force that shapes the symphony of rain.

Topography

The Earth's topography is the grand stage upon which precipitation patterns unfold, adorned with geographic nuances that add depth to the rainfall symphony. Mountains, valleys, and plains interact with air masses, influencing their ascent or descent and shaping the spatial distribution of rainfall.

Mountainous Terrain

- **Ascent and Enhanced Condensation:** Mountains play a pivotal role in the rainfall narrative. As moist air ascends a mountain range, it undergoes adiabatic cooling. The cooling process enhances condensation, transforming the ascending air mass into clouds. This phenomenon results in increased rainfall on the windward side of the mountain.

- **Leeward Side and the Rain Shadow Effect:** On the leeward side of the mountain, a contrasting scenario unfolds. As air descends, it undergoes adiabatic warming, creating conditions less favorable for condensation. This leeward side experiences a rain shadow effect, characterized by drier conditions and reduced rainfall.

Valleys and Plains

- **Influence on Air Mass Movements:** Valleys and plains, while not as topographically imposing as mountains, also influence rainfall patterns. They guide the movement of air masses, facilitating the ascent or descent that contributes to the spatial distribution of precipitation.

- **Interplay with Atmospheric Dynamics:** The interplay between topography and atmospheric dynamics creates a multifaceted stage for rainfall. The topographical features become integral components of the atmosphere, influencing the intensity and distribution of precipitation.

Air Currents

The movement of air currents across the globe shapes precipitation patterns with finesse. Trade winds, prevailing westerlies, and polar easterlies dictate the movement of air masses, influencing where precipitation occurs. Convergence zones, where air masses collide, become focal points for intense rainfall.

Trade Winds

- **Equatorial Convergence Zone:** Trade winds converge near the equator, creating the equatorial convergence zone. That becomes a breeding ground for intense rainfall. The warm, moist air rises, cools, and condenses, giving birth to the lush rainforests that characterize equatorial regions.

- **Tropical Rain Belts:** Trade winds, in their easterly course, also contribute to the formation of tropical rain belts. These bands of concentrated rainfall encircle the Earth, creating the climatic conditions that support diverse ecosystems.

Prevailing Westerlies and Polar Easterlies

- **Mid-Latitude Dynamics:** Prevailing westerlies dominate the mid-latitudes, and polar easterlies influence high latitudes, contributing to the mid-latitude dynamics of rainfall. These air currents guide weather systems, influencing precipitation patterns in temperate regions.

- **Storm Tracks and Frontal Boundaries:** The convergence of air masses along frontal boundaries, influenced by prevailing westerlies, becomes a theater for dynamic weather patterns. Storm tracks, shaped by these air currents, become corridors of intense rainfall.

Understanding these atmospheric currents unveils the intricate choreography of rain distribution on a planetary scale. The movements of air masses, driven by the Earth's rotation and solar heating, create a dynamic interplay that orchestrates rainfall in diverse regions across the globe.

Humidity

Humidity, a measure of the moisture content in the air, is a critical player in the precipitation narrative. It sets the rhythm for the dance of moisture, contributing to both the birth of clouds through evaporation and the eventual precipitation through condensation.

High Humidity Fosters Evaporation

- **Moisture-Laden Atmosphere:** High humidity levels create a moisture-laden atmosphere conducive to evaporation. Water bodies, soil, and vegetation release moisture into the air, saturating it with water vapor.

- **Evaporation from Oceans:** In regions with high humidity, like coastal areas and tropical climates, oceans play a significant role. The warm ocean surfaces provide ample moisture for evaporation, becoming the primary source for the moisture-rich air masses that fuel precipitation.

Condensation and Precipitation

- **Saturation and Condensation:** As air saturated with moisture ascends or encounters cooler conditions, it reaches its saturation point. That triggers the process of condensation, where water vapor transforms into tiny droplets or ice crystals, forming clouds.

- **Birth of Raindrops:** The condensed droplets, growing in size, become raindrops. The delicate balance between humidity and temperature determines when condensation prevails, leading to the birth of raindrops that will descend as precipitation.

Intensity and Duration of Rainfall

- **Humidity and Rainfall Intensity:** The intensity of rainfall is closely linked to humidity levels. High humidity contributes to more significant evaporation, creating the conditions for intense and prolonged rainfall events.

- **Seasonal Variations:** Humidity levels also exhibit seasonal variations, influencing the character of rainfall in different periods. Understanding these variations is crucial for deciphering the nuances of precipitation dynamics.

Deciphering the Symphony

In the grand symphony of rainfall, temperature fluctuations, topography, air currents, and humidity intertwine, creating a harmonious dance that sustains life on Earth. The interconnectedness of these factors forms a complex web, and deciphering their symphony provides insights into the diverse rainfall patterns witnessed across the globe.

The Interplay of Those Factors

- **Dynamic Relationships:** The relationship between temperature and humidity, the influence of topography on air masses, and the choreography of air currents contribute to the dynamic interplay that shapes rainfall patterns.

- **Regional Nuances:** Different regions experience unique combinations of these factors, giving rise to diverse climates and ecosystems. Each region tells a distinct rainfall story, from the monsoons in Southeast Asia influenced by oceanic and continental air masses to the temperate rainfall patterns shaped by prevailing westerlies.

- **Impact on Ecosystems:** The influence of these factors extends beyond meteorological dynamics to ecosystem health. Rainfall patterns dictate the availability of water resources, influencing the flora and fauna that thrive in specific regions.

As you delve into the factors influencing rainfall, you witness the intricate choreography of nature's ballet. From the nuanced guidance of temperature fluctuations to the dramatic topographical stage, the orchestrated movements of air currents, and the rhythmic interplay of humidity, each factor contributes to the symphony of rainfall.

Understanding this symphony is not merely an academic pursuit. It's a journey into the heart of Earth's vitality, where raindrops become the verses that narrate the story of life itself. In the ongoing narrative of the planet's water cycle, these factors continue to dance, creating the ever-changing melody of rainfall that sustains the beauty and diversity of this world.

Predicting Precipitation

In the ever-changing tapestry of this planet's climate, the ability to predict precipitation is paramount. It guides your preparedness for weather events, dictates agricultural practices, and helps people to understand Earth's water cycle. In this section, you'll journey through modern science and traditional wisdom, learning about the methods used to predict precipitation and bridge the gap between cutting-edge technology and ancestral insights.

Meteorological Forecasts

In modern science, meteorological forecasts are the guiding compass in anticipating precipitation patterns. Utilizing state-of-the-art technology, weather scientists harness the power of advanced tools to analyze vast datasets, interpret satellite imagery, and run sophisticated computer models. These tools enable them to predict atmospheric conditions, offering valuable insights into when and where precipitation will occur.

Advanced Technology at Play

- **Data Analysis:** Meteorologists delve into an extensive array of data, ranging from temperature and humidity levels to air pressure and wind patterns. Analyzing this data allows them to discern the complex interplay of factors that contribute to precipitation.

- **Satellite Imagery:** High-resolution satellite imagery provides a bird's-eye view of atmospheric conditions. It allows scientists to track cloud formations, identify weather systems, and monitor the development of potential precipitation events.

- **Computer Models:** Based on the collected data, advanced computer models simulate the atmosphere's behavior. These models consider various variables, enabling meteorologists to predict the timing, intensity, and duration of precipitation events.

Short-Term Forecasts to Extended Projections

- **Hourly and Daily Predictions:** Short-term forecasts, ranging from hourly to daily predictions, offer insights into imminent weather changes. These forecasts are crucial for planning daily activities, travel, and local events.

- **Extended Projections:** Meteorologists also provide extended projections that cover longer time frames, such as weekly or monthly forecasts. While these projections are uncertain, they offer valuable insights for mid-range planning and preparation.

Traditional Wisdom

Beyond cutting-edge technology, traditional wisdom cultivated over generations offers a unique perspective on predicting precipitation. Indigenous communities, deeply connected to the natural world, have developed a keen understanding of impending weather changes by observing natural indicators. This harmonious integration of ancestral knowledge with contemporary forecasting methods enriches the ability to foresee precipitation events.

Nature's Indicators

- **Animal Behavior:** Observing the behavior of animals has long been recognized as a reliable indicator of impending weather changes. Birds flying lower, cows lying down, or ants building their nests higher signal changes in atmospheric conditions.
- **Cloud Formations:** The art of reading cloud formations is a skill passed down through generations. Cloud types, colors, and patterns provide clues about upcoming weather. For example, the towering cumulonimbus clouds often herald thunderstorms.
- **Atmospheric Phenomena:** Natural occurrences like the halo around the moon or the red hues during sunrise and sunset have been observed for centuries as signs of changing weather. These atmospheric phenomena are woven into the fabric of traditional forecasting.

Ancestral Insights:

- **Cultural Knowledge:** Indigenous cultures often have specific cultural knowledge and rituals tied to weather predictions. This knowledge is shared within communities and plays a vital role in agricultural practices, hunting, and other aspects of daily life.
- **Interconnectedness with Nature:** Traditional forecasting emphasizes the interconnectedness between humans and nature. It recognizes that surroundings offer subtle cues about the changing rhythms of the natural world.

Rainfall Patterns and Climate Zones

Understanding the broader context of rainfall patterns in different climate zones contributes to more accurate predictions. Different regions exhibit distinct precipitation characteristics influenced by their proximity to the equator, local geography, and atmospheric dynamics. Recognizing these climatic nuances enhances the ability to predict when specific regions are more likely to experience rainfall.

Tropical Rainforests

Tropical regions near the equator experience consistent and heavy rainfall throughout the year.
https://www.pexels.com/photo/photo-of-foggy-forest-4633377/

- **Proximity to the Equator:** Tropical regions near the equator experience consistent and heavy rainfall throughout the year. The sun's direct rays at the equator create warm air, leading to the ascent of moist air masses and frequent precipitation.
- **Diverse Ecosystems:** The lush tropical rainforests are a testament to the abundance of rainfall. The consistent water supply supports diverse ecosystems, making accurate predictions crucial for managing these rich and fragile environments.

Arid and Semi-Arid Regions

- **Sporadic but Intense Precipitation:** Arid and semi-arid regions, like deserts, may experience sporadic but intense precipitation events. Understanding the factors contributing to these infrequent but impactful rainfall events is essential for water resource management.

- **Flash Flooding Risks:** In arid regions, the soil may have low permeability, leading to rapid runoff during intense rainfall. It poses the risk of flash flooding, making accurate predictions vital for mitigating potential hazards.

Temperate Climates

- **Seasonal Variations:** Temperate climates often exhibit distinct seasons with variations in precipitation. Understanding the seasonal patterns allows for better predictions regarding when rain is more likely to occur and its potential impact on agriculture and ecosystems.
- **Influence of Prevailing Winds:** Prevailing westerlies in temperate regions play a role in shaping rainfall patterns. Understanding the influence of these wind patterns contributes to accurate predictions.

Bridging the Gap

The synergy between modern science and traditional wisdom offers a holistic approach to predicting precipitation. While meteorological forecasts provide precise and data-driven predictions, traditional knowledge systems offer a nuanced understanding of nature's subtle cues. Integrating these knowledge systems enhances the ability to anticipate and adapt to changing weather conditions.

Cross-Cultural Collaboration

- **Knowledge Exchange:** Facilitating a cross-cultural exchange of meteorological knowledge enriches the collective understanding of weather patterns. Meteorologists can benefit from insights gained through traditional wisdom and vice versa.
- **Community Engagement:** Involving local communities in weather monitoring and prediction fosters a sense of ownership and empowerment. With their deep connection to the land, Indigenous communities contribute valuable observations that can complement scientific data.

Climate Resilience

- **Adaptive Strategies:** Incorporating traditional wisdom into climate resilience strategies enhances the adaptability of communities. Traditional forecasting methods, rooted in centuries of observation, offer early warnings and guide adaptive practices.

- **Preserving Biodiversity:** Accurate predictions are crucial for preserving biodiversity in various ecosystems. Indigenous knowledge, intimately tied to the rhythms of nature, contributes to sustainable practices that protect diverse flora and fauna.

Modern science and traditional wisdom play indispensable roles in the intricate dance of predicting precipitation. Meteorological forecasts, with their cutting-edge technology and data-driven precision, provide you with valuable insights into the complex dynamics of the atmosphere. Simultaneously, traditional knowledge systems, cultivated over generations, offer a profound connection to the natural world and its subtle indicators.

In this chapter, you've navigated the intricate realms of the water cycle, unveiling the science behind precipitation. From the ephemeral journey of water molecules through evaporation to the intricate factors influencing rainfall, the dance of precipitation is a symphony conducted by nature itself. As you explore the mechanisms shaping rainfall patterns, the stage is set for a deeper understanding of this planet's atmospheric dynamics.

Chapter 3: Picking Your Spot

In the pursuit of sustainable water practices, the art of harvesting rainwater stands as a pivotal solution. Yet, its success hinges on selecting the optimal spot. From the size and material of your roof to the lay of the land and the nuances of local climate, each factor plays a role in determining water catchment efficiency. This chapter considers picking the perfect spot, providing you with insights to balance functionality, aesthetics, and environmental impact.

Rooftop Considerations

In rainwater harvesting, the rooftop takes center stage. It's where the transformation of precipitation into a valuable resource begins. The dimensions, angles, and materials of your roof play a key role in determining the volume and quality of the rainwater you can capture.

In rainwater harvesting, the rooftop takes center stage.
https://www.pexels.com/photo/photo-of-roof-while-raining-2663254/

Roof Area and Collection Efficiency

The size of your roof directly dictates the potential volume of rainwater you will collect. This catchment area, a critical metric in rainwater harvesting, is determined by accurately measuring the dimensions of your roof. Precision ensures you harness the full potential of available rainfall.

Efficiency Considerations

While larger roofs offer more substantial catchment areas, efficient use of space and consideration for the intended use of harvested water is paramount.

- **Bigger Isn't Always Better:** Larger roofs indeed provide more significant catchment areas, enhancing the potential for water collection. However, it's crucial to strike a balance. Consider the available space, your water needs, and the intended use of the harvested water.

- **Tailoring to Needs:** Assess your water requirements and storage capacity. This understanding helps you optimize your catchment area to meet your specific needs without unnecessary excess.

Additional Considerations

Expanding on the calculation of catchment area, it's essential to consider factors that might affect efficiency:

- **Roof Slope Variation:** In cases where the roof has varying slopes, calculate the catchment area for each segment separately. This nuanced approach ensures accurate estimations.

- **Obstructions and Adjustments:** Account for any obstructions on the roof, such as chimneys or skylights, which may affect water flow. Adjustments to gutters and downspouts will optimize collection efficiency.

Roof Angles and Pitch

The angles and pitch of your roof add complexity to rainwater harvesting. They influence the speed of water runoff and the efficiency of collection systems.

Optimal Pitch

The pitch of your roof, its incline or slope, is a critical factor in rainwater harvesting.

- **Moderation Is Key:** A moderate pitch is often considered optimal for rainwater harvesting. Steep pitches lead to faster runoff, reducing the time water spends on the roof. Striking a balance is crucial for maximizing collection efficiency.
- **Preventing Runoff Issues:** A moderate pitch allows water to linger on the roof for a sufficient duration, promoting effective collection. It prevents issues associated with rapid runoff, ensuring a steady flow into your harvesting system.

Adjusting for Angles

Different roof designs and angles require tailored approaches to enhance water flow and collection efficiency.

- **Flat Roofs:** Flat roofs offer larger catchment areas but may require specialized systems to optimize water flow. Design adjustments and strategic placement of gutters will compensate for variations in roof angles.
- **Gabled Roofs:** Gabled roofs, with their slopes on either side, offer effective water runoff. Making sure that gutters are well-positioned to capture water along the slopes enhances efficiency.

Additional Considerations

- **Snow Load and Pitch:** In regions experiencing snowfall, consider the pitch's impact on snow accumulation. A steeper pitch will shed snow more effectively, preventing excessive amounts.
- **Roof Material Influence:** Certain roofing materials perform optimally at specific pitches. Investigate manufacturer recommendations to align your roof pitch with the chosen materials.

Roof Materials and Water Purity

The material composing your roof isn't merely an aesthetic choice. It's a key player in the quality of harvested rainwater. Different materials either introduce contaminants or contribute to cleaner water.

Metal Roofs

Corrosion-resistant metals like zinc or aluminum are popular choices for rainwater harvesting. They minimize leaching and contribute to cleaner water.

- **Durability and Purity:** Metal roofs are durable and corrosion-resistant, ensuring longevity. They also contribute to cleaner water by minimizing the introduction of contaminants.
- **Applicability to Harvesting Systems:** Metal roofs are compatible with various rainwater harvesting systems, offering versatility in design and implementation.

Asphalt Shingles

While common in roofing, asphalt shingles introduce small particles and contaminants into harvested water. Mitigating these concerns requires strategic solutions.

- **Particle Concerns:** Asphalt shingles shed small particles, affecting the purity of harvested water. Installing a first-flush diverter helps divert initial runoff, reducing particle content.
- **Regular Maintenance:** Periodic inspection and maintenance of asphalt roofs are crucial. Cleaning gutters and roof surfaces minimize the accumulation of debris and contaminants.

Treated Wood or Composite Shingles

These materials introduce chemicals into the harvested water, necessitating careful analysis and additional filtration measures.

- **Chemical Concerns:** Treated wood or composite shingles release chemicals into harvested water. Conducting thorough research on the specific materials used leads to awareness of potential contaminants.
- **Filtration Solutions:** Implementing additional filtration systems, such as sediment filters or activated carbon filters, will further purify water harvested from roofs with treated wood or composite shingles.

Additional Considerations:

- **Regular Roof Inspections:** Periodic inspections of your roof's condition are crucial. Detecting and addressing issues such as rust on metal roofs or deteriorating shingles on asphalt roofs ensures the longevity of the roof and the quality of the harvested water.

- **Material Longevity:** Consider the lifespan of roofing materials concerning your long-term rainwater harvesting goals. Investing in durable materials aligns with sustainability and reduces the frequency of replacements.

As you journey through rainwater harvesting, consider your rooftop as the conductor of a grand performance. The dimensions, angles, and materials harmonize to create a melody of efficient water collection. Precision in measurement, thoughtful consideration of pitch, and mindful selection of roofing materials contribute to the purity and abundance of the harvested rainwater. As you fine-tune each aspect, you optimize your rainwater harvesting system and contribute to the broader movement toward water conservation and a greener, more sustainable future.

Terrain and Topography

Beneath your feet lies a tapestry of contours and slopes. The terrain and topography of your property will either facilitate the smooth flow of water toward storage or present challenges that demand strategic solutions. It's time to explore how the natural features beneath your feet shape the intricate dance of rainwater harvesting.

Slope and Water Flow

The slope of your property is a dynamic force that dictates the natural flow of water. Efficiently harnessing this flow ensures rainwater travels from catchment surfaces to storage with minimal resistance.

- **Leveraging Natural Slopes:** Consider yourself fortunate if your property boasts natural slopes. Leverage these contours to guide water towards designated collection points. It will significantly reduce the need for complex drainage systems, allowing you to embrace the simplicity inherent in nature.

- **Creating Artificial Slopes:** In scenarios where natural slopes are insufficient or non-existent, consider introducing artificial slopes through strategic landscaping adjustments. By sculpting the terrain, you can redirect water flow, enhancing overall collection efficiency.

Water Flow Paths and Collection Points

Understanding how water moves across your property requires you to decipher nature's blueprint. Identifying optimal collection points involves a thoughtful analysis of the paths water takes during rainfall events.

- **Gutter Systems as Navigators**: Well-designed gutter systems act as the navigators of this natural journey. They guide water along predetermined paths, preventing chaotic runoff. Regular maintenance is the key to ensuring gutters remain clear, preventing blockages that could impede the smooth flow of water.

- **Strategic Placement of Storage Systems**: Positioning water storage tanks or reservoirs at points where runoff naturally converges is a masterstroke. It reduces the need for extensive piping systems, tapping into the simplicity of aligning with natural water courses.

- **Utilizing Swales and Berms**: Swales are depressions designed to redirect water flow, and berms can be strategically incorporated. These natural features assist in directing water toward desired collection points, enhancing the efficiency of rainwater harvesting.

Vegetation and Structures

The presence of vegetation on your property introduces both challenges and benefits to the rainwater harvesting narrative. Trees and plants act as natural filters, reducing contaminants in harvested water. However, they also contribute to debris, necessitating additional filtration measures.

- **Balancing Act of Nature**: Embrace the dual role of vegetation. While trees and plants contribute to water purity by acting as natural filters, they shed leaves and debris, potentially affecting the cleanliness of collected water. Striking a balance involves regular maintenance and additional filtration measures.

- **Strategic Planting for Water Retention**: Thoughtful vegetation placement will enhance water retention in the soil. It aids the prevention of soil erosion and promotes a more sustained release of water into the harvesting system.

Structures as Collection Points

Man-made structures, from sheds to outbuildings, significantly influence water flow patterns on your property. Thoughtful consideration of these elements will streamline the rainwater harvesting process.

- **Structures' Roofs as Catchment Areas:** Consider integrating existing structures into your rainwater harvesting system. Roofs of sheds or outbuildings serve as supplementary catchment areas, expanding the overall capacity for water collection.

- **Harmony of Aesthetics and Functionality:** Striking a balance between the aesthetic appeal of landscaping and the functional requirements of rainwater harvesting is an art. Thoughtful design ensures harmony between the natural environment and the infrastructure designed to capture and store rain's precious gift.

- **Utilizing Pervious Pavements:** Consider using permeable pavements in areas where hardscaping is unavoidable. These surfaces allow water to penetrate, reducing runoff and facilitating its absorption into the ground, contributing to the overall health of your rainwater harvesting system.

Navigating the Rainwater Harvesting Landscape

As you navigate the landscape of rainwater harvesting, the strategic placement of storage systems and the integration of both natural and built elements create a harmonious composition. Once perceived as static, the lay of the land beneath your feet now becomes a dynamic partner in the dance of water from the sky to storage.

Embrace the contours and slopes, work with the natural flow, and let your rainwater harvesting system become an extension of the landscape, seamlessly blending into the intricate design of your property. Nature and design, when choreographed with precision, transform rainwater harvesting from a practical necessity to a poetic interaction with the land itself. As you navigate this, remember that each slope, tree, and structure you've built contributes to a system that celebrates both the practicality and beauty of sustainable living.

Harmonizing Nature and Infrastructure

In exploring terrain and topography in rainwater harvesting, it's crucial to emphasize the profound connection between nature's dynamics and the man-made infrastructure designed to harness it. The process of rainwater harvesting unfolds most beautifully when these elements exist in harmony.

- **Adaptation to Local Conditions**: Recognize that each property is unique, and the strategies employed should be adapted to local topography, climate, and vegetation.

- **Continuous Observation and Adjustment:** As seasons change and landscapes evolve, ongoing observation and occasional adjustments to your rainwater harvesting setup ensure its continued effectiveness.

- **Educational Outreach:** Share your experiences and knowledge about terrain and topography in rainwater harvesting with your community. Encouraging sustainable practices contributes to a broader movement toward water conservation.

In this intricate dance of nature and design, your property becomes a canvas where rainwater transforms from a fleeting visitor into a cherished resident. Remember that the art of rainwater harvesting is not just about capturing water. It's about creating a sustainable and harmonious coexistence between human habitation and the natural world.

Regional and Climatic Factors

In rainwater harvesting, the final act transcends the microcosm of individual properties and delves into the macrocosm. Pay close attention to how local weather patterns and geographical features impact your harvesting decisions and potential yield. Understanding the broader climatic context will empower you to tailor your systems to regional intricacies, creating a symphony that harmonizes with nature's rhythms.

Proximity to Bodies of Water

The proximity of your location to bodies of water introduces a climatic nuance that profoundly influences rainwater availability.

Coastal Regions

Coastal regions embrace the ebb and flow of oceanic weather patterns, experiencing a unique dance with precipitation.

- **Consistency in Coastal Precipitation**: Coastal areas often enjoy more consistent rainfall, courtesy of the influence of oceanic weather patterns. This predictability enhances the reliability of rainwater harvesting systems, offering a steady water source.

- **Tailoring Systems for Coastal Reliability**: Individuals in coastal regions can fine-tune their rainwater harvesting setups with a degree of confidence in the regularity of precipitation. Focus shifts toward optimizing storage and efficiency of use.

Inland Areas

In contrast, inland locations face a different climatic cadence. The variability in rainfall patterns necessitates strategic considerations for ensuring a reliable water supply.

- **Navigating Variable Rainfall:** Inland areas experience more variable rainfall, demanding a flexible approach to rainwater harvesting. Supplemental water storage and efficient collection systems become crucial for maintaining a steady water supply amid fluctuations.

- **Adaptability as a Key Virtue**: The adaptability of rainwater harvesting systems in inland areas becomes a virtue. Solutions that accommodate the unpredictability of rainfall allow for a more resilient water strategy.

Local Weather Patterns

Understanding the unique weather patterns of your region is paramount. Different climates, including arid, tropical, and temperate, present distinct challenges and opportunities for rainwater harvesting.

Arid Climates

Harvested water becomes a precious resource in arid regions, where rainfall is sporadic but potentially intense.

- **Efficient Storage in Arid Realms**: Rainwater harvesting in arid climates demands efficient storage and utilization practices. Each drop must be cherished, making water conservation an inherent part of the harvesting strategy.

- **Microscale Weather Patterns:** Even within arid regions, microscale weather patterns influence the effectiveness of rainwater harvesting systems. Understanding local nuances allows for more precise system design, acknowledging the intricacies of arid climates.

Tropical Climates

Tropical regions, blessed with heavy and frequent rainfall, pose challenges to managing excess water during intense storms.

- **Managing Tropical Abundance:** While the abundance of rainfall in tropical climates is advantageous, managing excess water during intense storms becomes a necessary consideration. Efficient drainage systems and storage solutions are essential.

- **Navigating Tropical Microclimates:** Tropical regions often host diverse microclimates. Urban areas within tropical zones experience different rainfall patterns compared to rural or coastal areas, requiring nuanced system designs.

Temperate Climates

Temperate climates exhibit seasonal variations in rainfall, requiring adaptability in rainwater harvesting systems.

- **Year-Round Adaptability:** Adapting rainwater harvesting systems to seasonal variations ensures water availability in temperate climates year-round. Flexibility becomes the key to harnessing nature's changing moods.

- **Monitoring Seasonal Shifts:** Recognizing shifts in temperature and precipitation patterns during different seasons allows for proactive adjustments in rainwater harvesting systems. Continuous monitoring guarantees year-round effectiveness.

Microclimates and Microscale Weather Patterns

Even within a relatively small geographical area, microclimates and microscale weather patterns can vary, adding a layer of complexity to rainwater harvesting.

Urban Heat Islands

Urban areas create localized heat islands that influence weather patterns, impacting rainfall and temperature.

- **Microclimates in Urban Jungles:** Urban heat islands introduce microclimates that diverge from broader regional patterns. Understanding these nuances allows for more precise system design, acknowledging the intricacies of city life.

- **Balancing Urban Development and Rainwater Harvesting:** Microclimates are intensified in urban settings where concrete and asphalt dominate. Balancing the impermeability of urban surfaces with effective rainwater harvesting becomes crucial for sustainability.

Mountainous Terrain:

Mountainous regions experience orographic rainfall, influencing precipitation on windward and leeward sides.

https://www.pexels.com/photo/person-on-mountain-1647972/

Mountainous regions experience orographic rainfall, influencing precipitation on windward and leeward sides.

- **Navigating Mountain Dynamics:** In mountainous terrain, orographic rainfall leads to increased precipitation on windward sides and rain shadows on leeward sides. Strategic placement of harvesting systems considers these natural phenomena.

- **Harnessing Mountain Microclimates**: Microclimates within mountainous regions vary based on elevation, slope, and orientation. Understanding these intricacies aids in designing rainwater harvesting systems that align with the dynamic mountain environment.

Regulatory Considerations

Before finalizing your rainwater harvesting system, be aware of local regulations and guidelines. Regulatory considerations ensure a legal and sustainable approach to water collection.

Permitting and Regulations

Check if permits are required for rainwater harvesting systems. Some regions have regulations governing the size of storage tanks, runoff management, or water quality standards.

- **Navigating Legal Harmonies**: Understanding and complying with local permits and regulations ensures the legality and sustainability of your rainwater harvesting endeavors. Seek the necessary approvals to align your system with legal standards.
- **Educational Outreach on Regulatory Compliance**: Educate yourself and your community about the importance of adhering to regulations. Encourage awareness and compliance to foster a culture of legal and sustainable rainwater harvesting.

Community Guidelines

In communal living spaces or neighborhoods, adherence to community guidelines is essential. Collaborate with neighbors and local authorities to ensure your rainwater harvesting plans align with community standards.

- **Collective Responsibility**: Rainwater harvesting is not just an individual endeavor but a collective responsibility. Engage with your community to foster awareness and adherence to shared guidelines for sustainable water practices.
- **Sustainable Water Practices**: Work collaboratively with your community to establish guidelines that promote sustainable rainwater harvesting. Collective efforts enhance the effectiveness and acceptance of rainwater harvesting practices within the community.

Orchestrating Harmony in Rainwater Harvesting

As you conclude your exploration of regional and climatic factors in rainwater harvesting, envision it as orchestrating a harmonious symphony with nature. Tailoring your system to the unique climate of your region transforms rainwater harvesting from a utilitarian task to a poetic interaction with the environment.

- **Strategic Adaptation:** Strategic adaptation is the hallmark of a well-designed rainwater harvesting system. Whether in arid deserts, tropical paradises, or temperate havens, your system's ability to adapt ensures harmony with nature.

- **Educational Outreach:** Share your experiences and knowledge about regional and climatic factors in rainwater harvesting with your community. Fostering awareness and understanding ensures a broader movement toward sustainable water practices.

- **Continuous Monitoring:** Nature is dynamic, and so should be your approach. Continuous monitoring of weather patterns, system efficiency, and local regulations ensures that your rainwater harvesting system remains in tune with the evolving nature of your surroundings.

In the end, rainwater harvesting is all about harmonizing with the rhythms of nature. As you design and implement your system, let the regional and climatic factors become the notes in a melody that celebrates the beauty and sustainability of water stewardship.

Maximizing capture efficiency requires a nuanced understanding of your property's unique characteristics and consideration of local climate and regulations. You'll find the key to running a harmonious and effective rainwater harvesting system in this intricate dance between practicality, aesthetics, and personal preferences.

Embrace the challenge of selecting the perfect spot, for in this choice, you unlock the potential to transform raindrops into a sustainable source of life for your home and the environment.

Chapter 4: Designing Your Harvesting System

The journey of rainwater harvesting takes a pivotal turn in this chapter, where you discover the art and science of designing a harvesting system. Transforming the transient dance of raindrops into a sustainable water source requires thoughtful considerations.

Basic Components of a Harvesting System

In the intricate relationship between the heavens and the Earth, rainwater emerges as a precious elixir, a gift bestowed upon humanity by nature. From the initial pattern of raindrops on catchment surfaces to the safeguarding embrace of storage tanks, each component plays a crucial role in harmonizing with the environment.

Catchment Surfaces

At the heart of every rainwater harvesting system lies the catchment surface. Roofs stand as the primary catchment surfaces, whether shingled, metal, or tiled. Each material contributes unique characteristics that influence the purity and volume of the harvested water.

Roof Materials and Purity

The choice of roofing materials plays a crucial role in determining the quality of the collected rainwater. Different materials bring distinct advantages and considerations to the table:

- **Metal Roofs**: Known for their durability, metal roofs, composed of corrosion-resistant materials like zinc or aluminum, minimize the introduction of contaminants. It makes them an excellent choice for maintaining water purity.

Known for their durability, metal roofs, composed of corrosion-resistant materials like zinc or aluminum, minimize the introduction of contaminants.
Wikideas1, CC0, via Wikimedia Commons:
https://commons.wikimedia.org/wiki/File:Standing_seam_metal_roof_low_pitch_roof-3.jpg

- **Asphalt Shingles:** Common in residential structures, asphalt shingles are cost-effective but may introduce small particles and contaminants into the harvested water. Implementing a first-flush diverter will mitigate these concerns.

- **Tile or Concrete Roofs:** These materials offer durability and aesthetic appeal. However, their surfaces contribute to water hardness or introduce minerals. Filtration systems will be necessary for quality control.

Expanding Catchment Beyond Roofs

While roofs serve as primary catchment surfaces, thinking beyond conventional structures opens avenues for innovation in rainwater harvesting. Consider exploring:

- **Awnings:** Extend the reach of your catchment system by strategically placing awnings. They complement roof catchments and provide additional surfaces for rainwater collection.

- **Permeable Pavements**: Driveways and walkways made from permeable materials allow rainwater to penetrate the surface, contributing to the catchment. Incorporating these features will enhance the overall efficiency of your system.
- **Specially Designed Catchment Structures**: Innovative designs, such as catchment surfaces integrated into landscaping elements, will add both functionality and aesthetic appeal to your rainwater harvesting system.

Conveyance Systems

Once raindrops grace your catchment surface, the next act involves guiding them toward storage. Conveyance systems, comprising gutters and downspouts, conduct this liquid gold with efficiency and precision.

Gutter Systems

Well-designed gutter systems are the unsung heroes of rainwater harvesting. They ensure the smooth flow of water from the catchment surface to storage. Regular maintenance is essential to prevent blockages that compromise the efficiency of the entire conveyance system.

- **Material Considerations**: Choose gutter materials based on durability and compatibility with your catchment surface. Options include vinyl, aluminum, steel, and copper, each with its unique set of advantages.
- **Slope and Alignment:** Make sure that gutters are installed with a slight slope toward downspouts. Proper alignment prevents water stagnation and facilitates efficient drainage.

Downspouts and Diverter Systems

Downspouts act as conductors, guiding water from gutters to storage, while diverter systems enhance efficiency by preventing initial runoff, laden with debris and contaminants, from reaching storage directly.

- **Diverter Types:** First-flush diverters are crucial components. They redirect the initial runoff, which contains pollutants washed from the catchment surface, ensuring that only cleaner water enters the storage system.
- **Regular Maintenance:** Inspect and clean downspouts and diverter systems regularly to prevent clogs and blockages. This maintenance practice preserves the integrity and efficiency of your rainwater conveyance.

Filters

Before rainwater cascades into storage, it undergoes a refining process through filters that sift through impurities.

Mesh Screens

Basic mesh screens effectively capture larger debris, such as leaves and twigs, preventing them from entering the storage system. Regular cleaning is crucial to prevent clogging and maintain optimal filtration efficiency.

- **Maintenance Routine**: Include regular checks and cleaning of mesh screens in your rainwater harvesting maintenance routine. This simple step goes a long way toward preserving the functionality of your filtration system.
- **Cartridge Filters:** Cartridge filters become indispensable for finer filtration, especially in systems designed for potable water. These filters come in various micron ratings, allowing you to tailor filtration to the specific contaminants present in your region.
- **Micron Ratings**: Choose cartridge filters with appropriate micron ratings based on the quality of the harvested water and the contaminants you aim to remove. This precision ensures the purity of your collected rainwater.

Storage Tanks

The final destination for harvested rainwater is the storage tank. Tanks come in various materials, sizes, and shapes, each tailored to specific needs and space constraints.

Tank Materials

The choice of tank material influences durability, cost, and water quality. Consider the following options based on your preferences and specific requirements:

- **Polyethylene Tanks:** Lightweight and cost-effective polyethylene tanks are suitable for above-ground installations. They are resistant to corrosion and provide a practical solution for many applications.
- **Concrete Tanks:** Durable and suitable for underground installations, concrete tanks offer longevity and stability. However, proper sealing is essential to prevent minerals from leaching into the stored water.

- **Underground Cisterns:** Concealed beneath the ground, these tanks provide space-saving solutions. The choice of materials remains crucial to prevent contamination and ensure the purity of stored water.

Sizing Considerations

Calculating the ideal tank size involves considerations such as catchment area, average rainfall, and intended use. Oversized tanks ensure ample reserves for drier periods, offering a buffer against water scarcity.

- **Catchment Potential:** Evaluate the catchment potential of your surfaces, including roofs and additional catchment structures. This calculation forms the basis for determining the required storage capacity.
- **Average Rainfall:** Consider the average annual rainfall in your region. This data helps estimate the potential volume of harvested water, aiding in the selection of an appropriately sized storage tank.
- **Intended Use:** Define the purpose of your harvested water, whether for irrigation, household use, or potable water. Each use dictates the required volume, influencing the sizing of your storage tank.

Sustainability

Each component is fine-tuned by thoughtful design, from the catchment surfaces to the storage tanks.

- **Balancing Act:** Achieving a balance between functionality, aesthetics, and sustainability is key. Consider how each component contributes not only to the efficiency of your system but also to its overall impact on the environment.
- **Stewardship Responsibility:** Embrace the role of a responsible steward of water resources. The choices made in designing and implementing your rainwater harvesting system ripple through the broader ecosystem, reflecting a commitment to sustainability.
- **Continuous Care:** As you start to harness the rain's bounty, remember that continuous care is essential. Regular maintenance of components ensures the longevity and efficiency of your system, ensuring a seamless continuation of

the liquid magic.

In the next section, you'll learn about the intricate process of designing a rainwater harvesting system tailored to your specific needs and the unique characteristics of your environment.

Considerations Based on Intended Use

Rainwater harvesting transforms the art of utilizing rain into a versatile and sustainable resource. Each factor is carefully tuned to the intended use, whether nourishing the earth through irrigation, elevating daily household chores, or satisfying the thirst with potable water. It's time to dive deeper into each component, exploring the intricacies and considerations that make rainwater harvesting a personalized and sustainable endeavor.

Drip Irrigation Systems

Irrigation is a dance between water and soil. Here, precision and efficiency take center stage, with drip irrigation systems orchestrating a symphony of water droplets to nourish the earth. For those cultivating the earth, harvested rainwater becomes a lifeline for crops and greenery. Designing a system for irrigation requires considerations beyond purity, emphasizing volume and distribution efficiency. It's time to explore how this movement unfolds, ensuring that every drop fulfills its purpose in the grand composition of rainwater harvesting.

- **Efficient Water Distribution:** Drip irrigation systems minimize wastage with their ability to deliver water precisely to the plants' root zones. Coupled with appropriate filters, these systems ensure the efficient distribution of water, optimizing its use for agricultural purposes.

- **Sizing for Specific Needs:** Calculating the irrigation demand involves a nuanced understanding of the types of plants, soil characteristics, and local climate. A well-designed system aligns the availability of rainwater with the specific needs of the green space, promoting sustainable agricultural practices.

Household Filtration Systems

In a household context, rainwater elevates mundane chores to sustainable practices. Tailoring the system for household use involves addressing water quality and distribution for various domestic needs.

- **Enhancing Water Quality:** Household filtration systems play a pivotal role in ensuring water quality for domestic use. Using filters tailored to remove specific contaminants, such as sediment filters, activated carbon, or UV purification, enhances the purity of rainwater for everyday tasks.
- **Seamless Integration:** Designing the system to seamlessly integrate with household plumbing is crucial. Incorporating pressure pumps and distribution networks ensures a reliable supply for everyday tasks, transforming rainwater into a sustainable resource for daily living.

Drinking Water

For those venturing into potable rainwater, the design takes on a heightened level of precision. Rigorous filtration, disinfection, and compliance with health standards become paramount.

- **Multi-Barrier Filtration:** Employ advanced filtration systems with a multi-barrier approach. This may include sediment filtration, activated carbon, UV treatment, and, in some cases, reverse osmosis. Each layer contributes to the overall purity of harvested rainwater.
- **Continuous Monitoring:** Regularly testing the harvested water for contaminants is essential. Compliance with local health regulations ensures the potability of the water harvested, transforming rain into a safe and sustainable source of drinking water.

Scalability and Adaptability in Design

An effective rainwater harvesting system isn't just a static structure. It's a dynamic entity capable of growth. Scalability ensures that as needs evolve, the system expands to accommodate increased demand.

- **Oversizing Components:** Opt for oversized catchment surfaces and storage tanks. This provides a buffer for future expansion without requiring substantial redesign, allowing the system to grow in harmony with your evolving requirements.
- **Modular Conveyance Systems:** Design gutters and downspouts in a modular fashion. This allows for easy additions or modifications as catchment areas expand, facilitating seamless scalability without disrupting the existing structure.

Adaptability

Nature is dynamic, and so is the environment around your rainwater harvesting system. Designing with adaptability in mind ensures resilience against unforeseen changes and challenges.

- **Flexibility in Filtration**: Choose filter systems with modular components. This facilitates adjustments based on changes in water quality or the introduction of new contaminants, ensuring that the system can adapt to evolving conditions.

- **Weather-Responsive Controls**: Integrate weather-responsive controls for irrigation systems. This ensures adjustments based on rainfall forecasts, preventing overwatering during rainy periods. The adaptability to changing weather patterns makes the system responsive and efficient.

In the end, rainwater harvesting is all about harmonizing with the rhythms of nature. As you design and implement your system, let the intended use guide the composition, creating a masterpiece that transforms rain into a versatile and sustainable resource.

The Design Process

Rainwater harvesting is not just a pragmatic endeavor. It's a combination of thoughtful design, meticulous planning, and harmonious integration with nature. As you delve into the intricacies of the design process, envision it as composing a piece that resonates with the unique cadence of your environment. Here are the steps of this creative process, where each decision is a note in the melody of sustainability.

Step 1: Assessment of Catchment Potential

Evaluate Roof Characteristics

Assessing your roof's type, size, and material sets the architectural tone for your rainwater harvesting system. Each characteristic influences catchment potential and water quality.

- **Type:** Each type presents unique challenges and opportunities, from flat roofs to sloped designs. Assess how the architectural nuances impact water runoff and collection efficiency.

- **Size:** The size of your roof is a crucial factor in determining the catchment area. Larger roofs offer more potential for water collection but also demand careful considerations in system design.

- **Material:** The material of your roof goes beyond aesthetics. Different materials can introduce contaminants or enhance water purity. Consider corrosion-resistant metals or durable synthetic materials for optimal results.

Explore Additional Catchment Surfaces

Beyond the roof, additional catchment surfaces contribute to the richness of your water-harvesting efforts. A comprehensive assessment ensures optimal utilization of available surfaces.

- **Walls and Awnings:** Vertical surfaces like walls and awnings supplement your roof's catchment potential. Assess their contribution to overall water collection and factor them into your design.

- **Permeable Surfaces:** Evaluate permeable surfaces like driveways or courtyards. While these may not contribute directly to water catchment, understanding their role in water flow aids in efficient system design.

RAINWATER HARVESTING

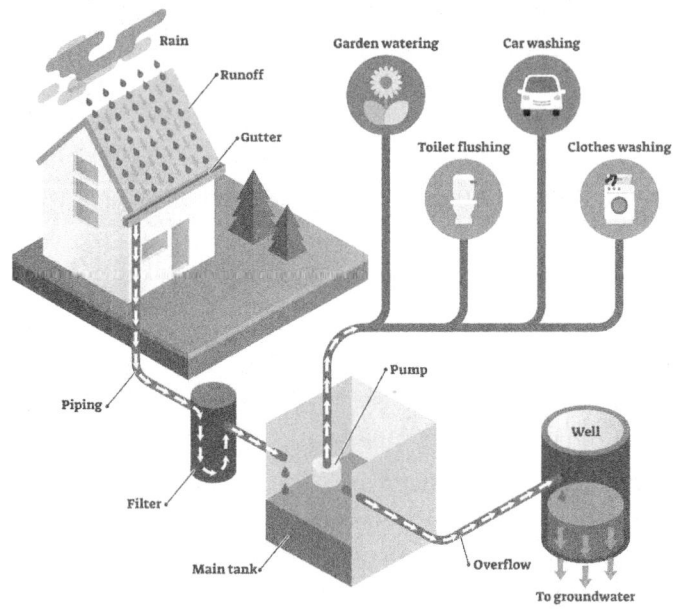

Beyond the roof, additional catchment surfaces contribute to the richness of your water-harvesting efforts.

Step 2: Calculating Water Demand

Determine Intended Use

Clearly defining the purpose of harvested water makes your goals clearer and your efforts focused. Each use dictates the required volume and quality for irrigation, household use, or potable water.

- **Irrigation:** If your focus is on irrigation, the demand may vary based on the types of plants and the size of the landscaped area. Understanding the specific water needs of your plants guides the design process.

- **Household Use:** For household use, consider daily activities like cooking, cleaning, and bathing. Clearly outlining the intended use ensures that your system aligns with practical water needs.

- **Potable Water:** If your goal is to harvest water for drinking, the highest purity standards are essential. The design must incorporate advanced filtration and purification components.

Calculate Demand

Based on the intended use, calculate the daily and seasonal water demand. This serves as the foundation for sizing components and designing the conveyance and storage systems.

- **Daily Demand:** Consider the daily water requirements for your chosen purpose. This includes understanding peak usage times and designing the system to meet these demands.

- **Seasonal Variations:** Recognize how water needs may vary throughout the seasons. Designing for seasonal fluctuations ensures a reliable supply of harvested water year-round.

Step 3: Selecting Appropriate Components

Roof and Catchment Design

Choosing roof materials and designing catchment surfaces is where aesthetics, durability, and efficiency come together harmoniously.

- **Aesthetics:** The visual appeal of your roof and catchment surfaces is integral to the overall design. Consider materials and designs that complement the architectural style of your property.

- **Durability:** Longevity is a key consideration. Select materials that withstand weathering and environmental factors, ensuring the sustained efficiency of your rainwater harvesting system.
- **Efficiency:** Striking the right balance between aesthetics and durability guarantees that your roof and catchment surfaces efficiently channel water to the collection system.

Conveyance Systems

Selecting gutter and downspout systems impacts the channels through which your water flows. Incorporate first-flush diverters and establish regular maintenance protocols.

- **Gutter Systems:** Choose gutter systems suitable for the catchment area. Consider factors like material, size, and shape to optimize water flow. Regular cleaning and maintenance prevent blockages.
- **Downspouts:** Efficient downspouts guide water downward from the roof to storage tanks. Position them strategically to maximize water capture and minimize runoff.
- **First-Flush Diverters:** Integrate first-flush diverters to minimize the initial runoff that may carry contaminants. This enhances the overall quality of harvested water.

Filtration Solutions

Choose appropriate filtration components based on water quality goals. Mesh screens, cartridge filters, or advanced purification systems ensure the desired water purity.

- **Mesh Screens:** These act as the first line of defense, preventing larger debris from entering the system. Regular cleaning maintains its effectiveness.
- **Cartridge Filters:** Mid-level filtration components that capture smaller particles. Choose cartridges based on your water quality objectives.
- **Advanced Purification:** Consider advanced purification systems like UV filters or reverse osmosis for potable water purposes. These ensure the highest level of water purity.

Storage Tanks

Consider tank materials and sizing based on catchment potential and water demand. Factor in scalability for future expansion, allowing your system to grow with changing needs.

- **Material Selection:** Choose materials that are durable, non-toxic, and corrosion-resistant. The most common include polyethylene, fiberglass, and concrete.

- **Sizing Considerations:** Your storage tank size should align with catchment potential and water demand. Calculate the necessary storage capacity to ensure a reliable water supply.

- **Scalability:** Opt for oversized storage tanks and consider modular tank designs. This accommodates future expansion without the need for significant redesign.

The size of your storage tanks should align with both catchment potential and water demand.

Step 4: System Integration and Distribution Planning

Integration with Existing Structures

Seamlessly integrate the rainwater harvesting system with existing structures. This includes plumbing for household use, irrigation networks, and potential expansion points.

- **Plumbing Integration:** Connect the rainwater harvesting system to existing plumbing for household use. Make sure that the harvested water seamlessly integrates with conventional water sources.

- **Irrigation Networks:** If the system is used for irrigation, plan for the integration of the rainwater supply with existing or new irrigation networks. Distribute water efficiently to landscaped areas.

Distribution Planning

Plan for efficient water distribution based on intended use. This may involve pressure pumps, drip irrigation networks, or household plumbing adjustments.

- **Pressure Pumps:** If needed, incorporate pressure pumps to ensure adequate water pressure for household use or irrigation. Proper distribution relies on maintaining consistent pressure.

- **Drip Irrigation Networks:** For irrigation purposes, design drip irrigation networks that deliver water directly to the base of plants. This conserves water and ensures targeted hydration.

- **Household Plumbing Adjustments:** If integrating with household use, plan for adjustments in plumbing to facilitate the seamless incorporation of harvested water into daily activities.

Step 5: Adaptability and Scalability Features

Modular Conveyance Additions

Design gutters and downspouts in a modular fashion. This allows for easy additions or modifications as catchment areas expand.

- **Modular Design:** Create a gutter and downspout system that can be easily extended or modified. This ensures adaptability as you expand your rainwater harvesting reach.

- **Future Catchment Areas:** Anticipate potential future catchment areas and design the system to accommodate these additions. This future-oriented approach prevents the need for significant overhauls.

Scalable Storage

Opt for oversized storage tanks and consider modular tank designs. This accommodates future expansion without the need for significant redesign.

- **Oversized Tanks:** Select storage tanks with a capacity that exceeds your current demand. This surplus capacity prepares your system for increased water needs in the future.

- **Modular Tank Designs:** Choose tank designs that allow for the easy addition of new modules. This scalable approach guarantees that your storage capacity can evolve with changing requirements.

Adaptable Filtration Systems

Choose filtration systems with modular components. This facilitates adjustments based on changes in water quality or the introduction of new contaminants.

- **Modular Filtration:** Choose filtration systems with interchangeable components. This allows you to upgrade or modify the system to address evolving water quality concerns.

- **Contaminant-Specific Filtration:** If the water source changes, such as increased sedimentation, choose filtration components specifically targeting the identified contaminants. This ensures continued water purity.

Step 6: Weather-Responsive Controls (Optional)

Implementing Smart Controls

For irrigation systems, consider weather-responsive controls. These systems adjust watering schedules based on real-time weather data, preventing overwatering during rainy periods.

- **Smart Controllers:** Incorporate weather-responsive controllers into your irrigation system. These controllers use real-time weather data to adjust watering schedules, optimizing water use.

- **Rain Sensors:** Integrate rain sensors that automatically suspend irrigation during rainfall. This ensures that harvested rainwater is not wasted and promotes water conservation.

Monitoring and Adjustment Protocols

Establish protocols to monitor system performance. Regular checks, especially after significant weather events, ensure optimal function and allow for adjustments as needed.

- **Routine Checks:** Schedule routine checks of the entire rainwater harvesting system. Inspect gutters, downspouts, filtration systems, and storage tanks to promptly identify and address any issues.

- **Post-Weather Event Checks:** After significant weather events, conduct thorough inspections. Heavy rainfall or storms may impact system components, and proactive checks prevent potential problems.

- **Adjustment Protocols:** Develop clear protocols for making adjustments to the system. Systematic adjustments maintain system efficiency, whether adapting to changing water quality or expanding the catchment area.

A Blueprint for Sustainable Water

Your rainwater harvesting system is a functional setup and a blueprint for sustainable water stewardship. Each decision, from catchment selection to storage tank sizing, becomes a stroke in the canvas of environmental responsibility.

- **Holistic Water Management:** Designing a rainwater harvesting system is a holistic approach to water management. It's a conscious choice to nurture nature's gift responsibly.

- **Adaptability for the Future:** As you face an ever-changing world, your rainwater harvesting system becomes a beacon of adaptability. Scalability and flexibility ensure it meets the challenges and opportunities of the future.

- **Educational Outreach:** Share your design insights with your community. Foster awareness and understanding of rainwater harvesting as a sustainable practice. Encourage others to embark on the journey of designing their systems.

The intricate design process of rainwater harvesting is a composition that harmonizes with the natural rhythms of your surroundings. Each decision, from selecting roof materials to integrating distribution networks, contributes to the seamless flow of your water-harvesting melody. Let your rainwater harvesting system be a testament to the artistry of sustainable living, where every drop is a note in the symphony of water stewardship.

Chapter 5: Storage Systems – Barrels, Gutters, and Tanks

Rainwater harvesting is a sustainable practice that conserves water and provides an eco-friendly alternative for various purposes. Choosing the right storage solution is paramount to the success of your rainwater harvesting venture. From the simplicity of gutters and barrels to the substantial presence of tanks, understanding the diverse storage options will empower you to make informed decisions that align with your needs, budget, and environmental considerations.

Short-Term Storage: Gutters and Barrels

Rainwater harvesting stands at the forefront of sustainable practices, offering a conscientious approach to water conservation. Gutters and barrels are two critical components of rainwater harvesting's initial stages. These short-term storage solutions play a pivotal role in collecting rainwater efficiently, providing an immediate and accessible source for a multitude of purposes. From materials and capacities to advantages and drawbacks, it's time to navigate the intricacies of these essential components.

Gutters

Gutters, the silent architects of rainwater harvesting, form the inaugural phase of this sustainable practice.

Gutters, the silent architects of rainwater harvesting, form the inaugural phase of this sustainable practice. As the first line of defense, gutters are crafted from materials such as aluminum, steel, or PVC. These unassuming channels, elegantly positioned along the roofline, play a crucial role in directing rainwater into downspouts, initiating the journey of collected water. Here's an exploration of the diverse materials, capacities, advantages, and drawbacks of gutters, unraveling the simplicity and effectiveness they bring to short-term rainwater storage.

Materials

Often crafted from materials like aluminum, steel, or PVC, gutters serve as the essential conduits that collect rainwater along the roofline, channeling it into downspouts. Each material brings its unique set of advantages to the rainwater harvesting ecosystem.

- **Aluminum:** Renowned for their lightweight nature and corrosion resistance, aluminum gutters are a preferred choice. Their durability and ease of handling make them practical and efficient solutions for homeowners seeking a reliable rainwater harvesting system.

- **Steel**: Recognized for robustness, steel gutters are a durable option. However, the susceptibility to rust requires regular maintenance to prevent deterioration over time, making them suitable for those willing to invest time in upkeep.
- **PVC**: As a cost-effective alternative, PVC gutters are resistant to both corrosion and rust. The versatility of PVC gutters is a standout feature, allowing them to adapt seamlessly to various roof types and shapes.

Capacities

The efficacy of gutters is closely tied to their size and the regional rainfall patterns. Regular maintenance, involving tasks such as clearing debris, is imperative to ensure optimal water flow and prevent overflow during heavy rainfall.

- **Size Variation**: Gutters come in various sizes, accommodating the diverse needs of different properties. Larger gutters handle more significant volumes of water, making them suitable for regions with higher rainfall.
- **Rainfall Considerations**: The capacity of gutters is directly influenced by the amount of rainfall in a specific area. Understanding regional rainfall patterns is crucial to determining the appropriate gutter size to collect and manage rainwater effectively.
- **Integration with Downspouts**: The seamless integration with downspouts also influences the capacity. Properly designed systems ensure water efficiently travels from the gutters to the downspouts, preventing overflow and maximizing storage.
- **Maintenance Impact**: Regular maintenance, such as clearing leaves and debris, is paramount for ensuring optimal water flow. Well-maintained gutters effectively manage higher capacities without the risk of clogs or overflowing.

Advantages

Gutters boast several advantages, making them an attractive choice for short-term rainwater storage:

- **Cost-Effective**: Gutters present a budget-friendly option, democratizing rainwater harvesting by making it accessible to a wide spectrum of homeowners.

- **Ease of Installation:** The simplicity of gutter systems translates into easy installation, often making them a popular choice for those who enjoy engaging in DIY projects around their homes.
- **Versatility:** The adaptability of gutters to different roof types and shapes adds to their appeal. This versatility makes them suitable for a variety of architectural designs.

Drawbacks

- **Limited Storage Capacity:** Gutters are not designed for extensive water storage, rendering them more suitable for immediate use rather than long-term storage solutions.
- **Maintenance Requirements:** Frequent cleaning is necessary to prevent clogs and overflow, demanding regular homeowner attention. While the maintenance is straightforward, it's an ongoing commitment.

Barrels

As you move along the rainwater harvesting journey, barrels step onto the stage, providing an elegant enhancement to short-term storage solutions. Positioned strategically under downspouts, these unassuming containers elevate the rainwater collection, offering functionality and an aesthetic touch to the process. In this section, you'll discover rain barrels' various materials, capacities, advantages, and drawbacks, unraveling their role in enhancing short-term rainwater storage.

Materials

Rain barrels strategically positioned under downspouts complement gutters in collecting rainwater. These barrels come in various materials, each with unique merits, adding a layer of customization to the rainwater harvesting experience.

- **Plastic:** Lightweight and corrosion-resistant, plastic barrels are a popular and practical choice. Their ease of handling and suitability for various climates make them a go-to option for many homeowners.
- **Wood:** For those looking for a touch of rustic aesthetics, wooden barrels fit the bill. While they add an attractive element to the garden landscape, they require more maintenance to preserve their charm.
- **Metal:** Known for their durability, metal barrels are sturdy and chosen for their longevity. The trade-off is their susceptibility to

rust, requiring you to weigh the benefits against potential maintenance needs.

Capacities

Rain barrels offer a range of capacities, typically 50 to 100 gallons. This variability allows you to select a size that aligns with your water needs and the available space on your property.

- **Size Options:** Rain barrels come in various sizes to cater to different usage requirements. Smaller barrels are suitable for limited-space areas, while larger ones accommodate higher water demands.

- **Modularity:** You can install multiple barrels in a modular fashion, creating a collective storage system with increased capacity. This modularity provides flexibility in adapting to changing water consumption patterns.

- **Customization:** Some barrels are designed with customizable capacities, allowing you to choose the size that best fits your specific needs. This customization ensures that the rain barrel aligns seamlessly with the property's requirements.

- **Overflow Prevention Features:** Many rain barrels incorporate features like overflow valves or outlets to manage higher capacities effectively. These mechanisms ensure excess water is directed away from the barrel, preventing overflow and potential water wastage.

Advantages

- **Affordability:** Rain barrels are cost-effective, align with budget considerations, and make rainwater harvesting accessible to a broad audience.

- **Easy Installation:** Like gutters, rain barrels are relatively easy to install, often making them popular for DIY projects. This simplicity adds to their appeal, especially for those with a hands-on approach to home improvement.

- **Immediate Access:** Rain barrels offer immediate access to harvested rainwater, facilitating activities like watering plants or washing outdoor surfaces without delay.

Drawbacks

- **Limited Storage Capacity:** Like gutters, rain barrels are not designed for extensive storage. They are best suited for short-term use, emphasizing immediate accessibility over prolonged storage needs.

• **Maintenance Requirements:** Regular cleaning and filtering are necessary to ensure water quality and prevent issues such as mosquito breeding. While maintenance is crucial, it is a manageable aspect for those committed to reaping the benefits of rainwater harvesting.

Understanding the materials, capacities, advantages, and drawbacks of gutters and barrels is pivotal for making informed decisions. Homeowners looking to embark on a rainwater harvesting journey can blend the efficiency of gutters with the accessibility of barrels to create a well-rounded and sustainable system. By embracing these short-term solutions, you contribute to your water conservation efforts and the broader movement toward environmentally conscious living.

Making an informed decision about short-term rainwater harvesting solutions requires thoughtful consideration of individual needs, property characteristics, and commitment to maintenance. The synergy of gutters and barrels provides a balanced approach, offering efficiency and accessibility to homeowners seeking to integrate sustainability into their daily lives.

Long-Term Storage: Tanks

In the quest for sustainable water management, long-term rainwater harvesting solutions become paramount, especially for those facing infrequent rainfall or high water demand scenarios. Tanks emerge as stalwart players in this arena, offering substantial storage capacities to meet the needs of residential, commercial, and agricultural applications.

Materials

Tanks, the cornerstone of long-term rainwater harvesting, are crafted from a variety of materials, each presenting a unique set of characteristics.

• **Polyethylene Tanks:** Lightweight and resistant to corrosion, polyethylene tanks offer a practical solution for those seeking durability without the burden of excessive weight. Their versatility extends to above-ground installations, making them accessible for various applications.

- **Fiberglass Tanks**: Renowned for their durability, fiberglass tanks are a robust choice, particularly suitable for underground installation. This feature preserves property aesthetics and optimizes space usage. Fiberglass tanks are resistant to corrosion, making them a reliable long-term option.

- **Concrete Tanks**: Robust and sturdy concrete tanks are known for their durability. However, their weight is a limiting factor, and they are typically employed in scenarios where above-ground installation is feasible. Concrete tanks provide a solid and long-lasting solution for extensive rainwater storage needs.

- **Steel Tanks**: Sturdy and capable of withstanding external pressures, steel tanks are a common choice for above-ground installations. However, they are prone to rust, necessitating careful consideration of maintenance practices to ensure their longevity.

Capacities

The allure of tanks lies in their ability to cater to a broad spectrum of water storage needs, from modest residential applications to large-scale commercial and agricultural requirements.

- **Range of Capacities**: Tanks offer a diverse range of capacities, accommodating the specific demands of various users. From a few hundred gallons to several thousand, the flexibility in size ensures that individuals and businesses can tailor their rainwater harvesting systems to their unique requirements.

- **Residential Applications**: Smaller tank capacities are often suitable for residential applications, providing homeowners with a reliable and sustainable water source for domestic use, landscaping, and other household needs.

- **Commercial and Agricultural Needs**: Larger tank capacities find their niche in commercial and agricultural settings where water demand is more substantial. Tanks are integral to ensuring a consistent and sufficient water supply for crops, livestock, and industrial processes.

Advantages

Tanks bring many advantages to the table, making them indispensable for those seeking robust, long-term rainwater harvesting solutions.

- **Substantial Storage Capacity:** The primary strength of tanks lies in their ability to store significant volumes of water, making them ideal for regions with infrequent rainfall or areas facing high water demand. This feature gives you a reliable and consistent water supply even during dry spells.

- **Customization for Underground Installation:** Tanks can be customized for underground installation. They're a particularly valuable option for those aiming to preserve above-ground space or maintain property aesthetics. This underground configuration enhances space efficiency and protects the tanks from external elements.

- **Versatility in Applications:** Tanks cater to a wide array of applications, from residential water conservation to large-scale agricultural and commercial operations. Their adaptability positions them as versatile solutions for diverse water storage needs.

Drawbacks

While tanks offer substantial benefits, it's essential to acknowledge the considerations and challenges associated with implementing them.

- **Higher Upfront Costs:** Compared to short-term solutions like gutters and barrels, tanks come with higher upfront costs. The investment required to purchase and install tanks is a significant consideration for those managing budget constraints.

- **Professional Installation Required:** Installing tanks, especially in customized or underground configurations, often requires professional assistance. It adds to the overall costs and underscores the importance of ensuring the installation is carried out precisely.

- **Regular Maintenance is Crucial:** Regular maintenance is imperative to ensure tank longevity and optimal functionality. It includes checking for corrosion, cleaning, and addressing any potential issues promptly. Ignoring maintenance will lead to deterioration and reduce the lifespan of the system.

Implementing Tanks

As you navigate long-term rainwater harvesting, implementing tanks emerges as a pivotal step toward building water resilience. Understanding tanks' materials, capacities, advantages, and drawbacks equips individuals and businesses to make informed decisions aligning with their specific needs and circumstances.

- **Material Selection:** Choosing the right material for a tank involves a delicate balance between durability, weight considerations, and the intended application. Polyethylene tanks, with their lightweight nature and corrosion resistance, are ideal for above-ground residential installations. Fiberglass tanks, with their durability and suitability for underground use, offer a discreet solution that doesn't compromise property aesthetics. Concrete tanks, though heavier, provide robustness for various applications, while steel tanks, which are also sturdy, require diligent maintenance to combat rust.

- **Calculating Capacities:** Determining the appropriate tank capacity depends on accurately assessing water demand. Residential users may find smaller capacities sufficient for daily needs, while those engaged in agriculture or commercial activities require larger tanks for a consistent and reliable water supply. Understanding regional rainfall patterns and the frequency of dry spells aids in fine-tuning the capacity to match actual needs.

- **Advantages for Varied Applications:** The versatility of tanks shines in their ability to cater to a spectrum of applications. In a residential setting, tanks are a sustainable water source for daily activities, landscape irrigation, and emergency preparedness. For commercial enterprises, tanks provide a reliable supply for industrial processes, reducing dependence on external water sources. In agriculture, where water is a lifeline for crops and livestock, tanks guarantee a consistent supply, contributing to sustainability and productivity.

- **Mitigating Drawbacks:** While the drawbacks associated with tanks are notable, proactive planning mitigates potential challenges. Addressing the higher upfront costs involves considering the long-term benefits and return on investment that tanks bring. Seeking professional assistance during installation ensures the system is set up correctly, maximizing its efficiency and lifespan. Finally, regular maintenance should be viewed as a proactive investment rather than a reactive necessity, safeguarding the longevity and functionality of the rainwater harvesting system.

Using tanks in a rainwater harvesting system is not just a practical choice. It's a commitment to sustainable water practices. As individuals, communities, and businesses strive to reduce their environmental footprint, the role of tanks in water conservation becomes increasingly pivotal. The investment in tanks transcends the mere procurement of a storage system. It symbolizes a dedication to responsible water management and a proactive stance in securing water resources for the future.

As you integrate tanks into your rainwater harvesting endeavors, you contribute to personal water security and the broader movement promoting water resilience. Adopting long-term solutions like tanks represents a collective step towards a more sustainable and water-conscious future, where every drop is valued, conserved, and utilized precisely.

Maintenance and Integration

A well-maintained rainwater harvesting system is not just a reservoir for environmental consciousness but an investment in efficiency and longevity. Whether your chosen storage option is barrels, tanks, or a combination of both, routine maintenance is the key to its effectiveness. It's time to dive into the specifics to ensure your system runs like a finely tuned instrument.

Gutters

1. **Regular Cleaning:** The first line of defense in rainwater harvesting is your gutters. These conduits direct rain from your roof into the storage system. Regular cleaning is paramount to prevent clogs and ensure a smooth flow of water. The flow is restricted if clogged, just like cholesterol blocking blood vessels. Regular cleaning keeps the path clear and fluid.

2. **Inspection and Repair:** Take a closer look at your gutters periodically. Are they sagging or damaged? Repair any issues promptly to maintain their structural integrity. Gutters, much like a well-tuned instrument, need occasional adjustments. Tighten loose screws or replace damaged sections to keep everything in harmony.

3. **Leak Checks:** The connections between gutters and downspouts are potential leak points. Regularly inspect these junctions and fix

any leaks promptly. A small leak will disrupt the entire performance.

Barrels

1. **Interior Cleanliness:** The interior of your barrels is a breeding ground for potential issues, especially algae growth. Regular cleaning prevents these unwanted green guests.

2. **Crack and Leak Inspection:** Barrels, like any container, are susceptible to cracks and leaks. Periodically inspect them and promptly repair any damages. Just as a musician checks their instrument for cracks or warps, you must inspect your barrels. A small crack may seem insignificant, but it leads to a loss of the precious liquid.

3. **Securing the Lid:** The lid of your barrel is its first line of defense against debris and contamination. Ensure it's securely in place to maintain the purity of your harvested rainwater.

Tanks

1. **Structural Inspections:** Being larger structures, tanks require periodic inspections for leaks or structural damage. Catching these issues early on prevents more extensive problems. A tank with structural damage is a building with a compromised foundation. Regular inspections ensure everything stands strong.

2. **Interior Cleanliness:** Sediment buildup in tanks reduces their efficiency. Regularly clean the tank's interior to prevent this, ensuring a clear and unobstructed space for water storage. Routine cleaning keeps everything in top condition.

3. **Filtration System Monitoring:** If your tank has a filtration system, monitor it regularly for optimal performance. Clean or replace filters as needed. Filters in a rainwater harvesting system are like the strings on a guitar. Regular tuning (cleaning or replacement) ensures they produce the desired melody.

Harmonizing Functionality and Aesthetics

1. **Strategic Placement:** Consider the strategic placement of your storage systems. Optimize water collection while minimizing the visual impact on your property. Placing tanks strategically is like arranging furniture in a room. It must be functional (providing seating or storage) while not overpowering the visual space.

2. **Aesthetics**: Choose storage options that complement your property's aesthetic. Whether it's the sleek lines of a tank or the rustic charm of barrels, make sure it aligns with the visual theme. It should enhance, not detract, from the overall aesthetic of your property.

3. **Landscaping:** Integrate storage systems into your landscaping design. Treat them as functional elements within the overall aesthetic, harmonizing nature and function. Your storage system is a part of a garden orchestra. Each element plays its role, contributing to a beautiful design.

4. **Accessibility:** Make sure your storage systems are easily accessible for maintenance. However, consider the visual aspects to maintain the property's appeal. It must be practical for maintenance but shouldn't distract from the property aesthetics.

The harmonious interplay of maintenance and integration is the key to a successful rainwater harvesting system. Regular upkeep ensures the efficient functioning of your system, while thoughtful integration enhances the visual appeal of your property. Like a skilled conductor guiding an orchestra, you direct the elements of your rainwater harvesting system to create a synchronization between sustainability and beauty.

As you conclude this exploration of storage systems in rainwater harvesting, it's evident that each component plays a distinct role, contributing to the overall harmony of sustainable water management. Whether it's the simplicity of gutters guiding the initial flow, the charm of barrels storing raindrops like precious notes, or the grandeur of tanks conducting a monumental tune, the choice of storage systems defines the rhythm of this water-harvesting composition.

Chapter 6: Safety and Filtration – Ensuring Clean Water for Every Use

Rainwater harvesting demands the assurance of clean and safe water for everyday use. This chapter focuses on the meticulous process of safeguarding harvested rainwater, addressing potential contaminants, exploring filtration techniques, and delving into post-storage treatments to guarantee water safety. From understanding the sources of contamination to implementing effective filtration methods and post-filtration treatments, this chapter will guide you through the comprehensive measures necessary for maintaining consistently clean water.

Rainwater harvesting demands the assurance of clean and safe water for everyday use.
https://pixabay.com/vectors/virus-boat-doctor-team-rescue-7341187/

Potential Contaminants in Harvested Rainwater

In the pursuit of sustainable water practices, harvested rainwater often emerges as a promising source. However, the apparent purity of rainwater can be deceiving. Before thinking about water safety, you need to understand the potential contaminants that infiltrate harvested rainwater. These contaminants, originating from various sources, span a spectrum that includes atmospheric pollutants, debris, and biological agents, forming a complex tapestry that requires careful consideration.

Atmospheric Pollutants: Unseen Impurities Descending from the Skies

On its descent, rainwater encounters a myriad of atmospheric pollutants that compromise its purity. These pollutants, though invisible to the naked eye, significantly impact the quality of harvested rainwater.

1. **Airborne Particulate Matter:** The seemingly innocent act of rain falling from the sky brings with it the accumulation of airborne particulate matter. Dust, pollen, and other microscopic particles suspended in the air settle on rooftops and are inevitably carried into harvested rainwater. The rooftops, once pristine, become catchment areas for these impurities, introducing a layer of complexity to the water harvesting process.

   - **Impact on Water Quality:** While these particles might seem inconsequential individually, their cumulative presence affects the taste and visual clarity of harvested rainwater. Moreover, they contribute to clogging in filtration systems if not addressed.

   - **Clogging Concerns:** Accumulation of airborne particles leads to clogging in downspouts and gutters, affecting the efficiency of rainwater collection.

   - **Visual Clarity:** The presence of particulate matter results in cloudy or turbid water, affecting the visual aesthetics of harvested rainwater.

2. **Chemical Pollutants:** The industrial landscape, a significant contributor to air pollution, casts a subtle but impactful shadow on harvested rainwater. Industrial emissions release a cocktail of chemicals into the atmosphere, some of which find their way into rainwater during its descent.

- **Sulfur Compounds:** Factories and industrial facilities emit sulfur compounds that dissolve in rainwater, potentially leading to the formation of acid rain. The presence of sulfur compounds in harvested rainwater alters its pH and introduces acidity.

- **Heavy Metals:** The insidious nature of heavy metals, such as lead, mercury, and cadmium, is exacerbated by their release into the air through industrial processes. Once airborne, these metals settle on surfaces, including rooftops, becoming unwelcome guests in harvested rainwater.

- **Impact on Water Quality:** Chemical pollutants introduce a range of undesirable characteristics to rainwater, from altered taste and color to potential health hazards associated with heavy metal ingestion.

- **Health Considerations:** Consumption of rainwater contaminated with heavy metals poses serious health risks, emphasizing the importance of effective filtration.

3. **Microorganisms:** The atmosphere, though vast and seemingly pure, is home to a myriad of microorganisms that hitch a ride with raindrops. Bacteria, viruses, and fungi present in the air find their way into harvested rainwater during its journey from the clouds to catchment surfaces.

- **Impact on Water Quality:** While rainwater is generally considered free of harmful microorganisms, the potential introduction of these entities from the atmosphere emphasizes the need for thorough filtration and disinfection processes.

- **Filtration Challenges:** Addressing microorganisms requires specialized filtration methods to ensure the removal of potential pathogens.

- **Disinfection Considerations:** Microbial contamination highlights the importance of post-filtration disinfection to ensure the safety of harvested rainwater for various uses.

Debris and Environmental Factors: Ground-Level Challenges

Beyond atmospheric pollutants, ground-level factors contribute significantly to the potential contamination of harvested rainwater. These factors encompass a range of environmental challenges that merit attention.

1. **Roof Material Contamination:** The type of material used for roofing plays a pivotal role in determining the quality of harvested rainwater. Roof materials leach substances that compromise water purity.

   - **Asbestos Contamination:** In older buildings, roofs made of asbestos release fibers into rainwater, posing health risks if consumed.

   - **Treated Wood:** Roofs constructed with treated wood introduce chemicals into harvested rainwater, adding another layer of complexity to safety considerations.

   - **Impact on Water Quality:** Roof material contamination underscores the importance of selecting roofing materials carefully, especially when aiming to harvest rainwater for potable use.

   - **Material Selection:** Choosing roofing materials with minimal leaching properties is crucial for maintaining water quality.

   - **Health Awareness:** Educating yourself about the potential risks associated with specific roofing materials promotes informed decision-making.

2. **Overhanging Trees:** The charm of overhanging trees comes with an unintended consequence for rainwater harvesting. Leaves, bird droppings, and other organic matter from trees become potential contaminants.

   - **Leaves:** Falling leaves introduce organic matter that decomposes in harvested rainwater, impacting its quality.

   - **Bird Droppings:** Bird droppings, while seemingly innocuous, harbor bacteria and contribute to microbial contamination.

   - **Impact on Water Quality:** The natural beauty of overhanging trees brings with it the responsibility of managing potential contaminants, requiring proactive measures to ensure water purity.

   - **Regular Pruning:** Trimming overhanging branches reduces the likelihood of leaves and debris entering the rainwater harvesting system.

   - **Bird Deterrents:** Implementing measures to deter birds, such as installing bird spikes or nets, minimizes the introduction of bird-related contaminants.

3. **Animal Activity**: Birds, insects, and small animals find rooftops and catchment areas attractive, contributing to the potential contamination of harvested rainwater.

- **Birds:** Besides droppings, birds bring in feathers, nesting materials, and even small prey, all of which impact water quality.

- **Insects:** Insects, attracted to the moisture on rooftops, inadvertently become part of the harvested rainwater.

- **Impact on Water Quality:** Managing the presence of animals on rooftops is essential to prevent their contribution to contaminants, emphasizing the need for protective measures.

- **Mesh Barriers:** Installing mesh barriers or screens over gutters and downspouts prevents insects and larger debris from entering the system.

- **Bird-Proofing:** Employing bird-proofing measures, such as installing deterrents or netting, reduces the likelihood of bird-related contamination.

4. **Runoff from Surfaces**: Adjacent surfaces, such as driveways or areas with polluted soil, contribute to runoff that finds its way into the rainwater harvesting system.

- **Chemical Runoff:** Polluted soil or chemical-laden surfaces introduce substances into harvested rainwater.

- **Sediment and Debris:** Runoff carries sediment and debris, adding to the challenges of water quality maintenance.

- **Impact on Water Quality**: Managing runoff is a critical aspect of rainwater harvesting, requiring careful planning and measures to prevent the introduction of external contaminants.

- **Permeable Surfaces:** Implementing permeable surfaces in the vicinity reduces runoff and minimizes the influx of external contaminants.

- **Rain Gardens:** Designing rain gardens or buffer zones absorbs and filters runoff before it reaches the rainwater harvesting system.

As you unravel the intricate web of potential contaminants in harvested rainwater, it becomes evident that ensuring water purity is a multi-faceted challenge. From atmospheric pollutants descending from the skies to ground-level factors contributing to runoff, each element requires careful consideration.

Harvesting rainwater holds incredible potential for sustainable water practices, but this potential can only be fully realized with a nuanced understanding of the challenges at hand. Addressing potential contaminants involves a combination of thoughtful infrastructure design, regular maintenance practices, and a commitment to ongoing water quality monitoring.

Filtration Techniques and Their Efficacy

Water safety involves navigating the intricate world of filtration techniques. Filtration is not a one-size-fits-all solution. It's a nuanced and multi-faceted process designed to address specific types of contaminants. From mesh filters that act as the first line of defense to sophisticated reverse osmosis systems capable of removing a broad spectrum of impurities, each filtration method plays a unique role in ensuring the purity of harvested rainwater.

Mesh Filters

1. **Mechanism:** Mesh filters, typically crafted from materials like stainless steel or nylon, operate on a simple yet effective principle. They physically block larger particles and debris from entering the water system.

   - **Mesh Material:** The choice of materials, such as stainless steel or nylon, ensures durability and resilience against environmental factors.

   - **Mesh Pore Size:** Variations in mesh pore size allow customization based on the size of particles to be filtered.

2. **Optimal Applications:** These filters find their sweet spot in scenarios where the focus is on removing larger particles. They serve as ideal pre-filters, stationed strategically to protect subsequent filtration stages from potential clogging.

   - **Pre-Filtration:** Commonly employed as the first layer of defense, mesh filters prevent leaves, insects, and larger particulate matter from progressing further into the filtration system.

- **Rainwater Collection:** Mesh filters are integral to rainwater collection systems, ensuring that only clean water enters storage tanks.

3. **Efficacy:** While mesh filters exhibit high efficacy in trapping larger particles, their effectiveness wanes when it comes to dealing with dissolved pollutants or microorganisms.

- **Limitations:** Mesh filters aren't the silver bullet for contaminants at the molecular or microbial level, necessitating additional filtration stages.

- **Maintenance:** Regular cleaning is essential to prevent clogging and maintain optimal performance.

- **Replacement Schedule:** Periodic replacement of mesh filters ensures continued effectiveness, especially in areas with high debris levels.

Sediment Filters: Navigating the Finer Particles

1. **Mechanism:** Sediment filters employ materials like sand or fabric to trap finer particles suspended in the water, offering a more nuanced approach to filtration.

- **Material Variety:** The use of different materials provides flexibility to address specific particle sizes effectively.

- **Depth Filtration:** Some sediment filters utilize depth filtration, enhancing their capacity to capture particles throughout the entire filter depth.

2. **Optimal Applications:** Effective in removing smaller sediment particles, silt, and fine debris, sediment filters carve a niche in scenarios where precision in particle removal is crucial.

- **Fine Particle Removal:** Ideal for applications where the presence of fine sediment poses a challenge to water quality.

- **Pre-RO Filtration:** Sediment filters often serve as precursors to more advanced filtration methods like reverse osmosis.

3. **Efficacy:** Sediment filters provide good filtration for sediments but aren't as adept at handling dissolved pollutants or microorganisms.

- **Considerations**: You should be aware that sediment filters may need additional support to address contaminants beyond particulate matter.

- **Maintenance**: Regular monitoring and replacement of sediment filters are necessary to prevent clogging and maintain efficiency.

Activated Carbon Filters

1. **Mechanism**: Activated carbon filters work by adsorbing organic compounds, chemicals, and some gases from the water, making them versatile tools in the filtration arsenal.

   - **Adsorption Power**: The porous structure of activated carbon enhances its ability to attract and trap impurities effectively.

   - **Microporous Structure**: Activated carbon possesses a microporous structure, providing a large surface area for adsorption.

2. **Optimal Applications**: Activated carbon filters shine in applications where the focus is on removing organic contaminants, chlorine, and specific chemicals that affect taste and odor.

   - **Organic Contaminant Removal**: Well-suited for scenarios where the water source is prone to organic impurities.

   - **Taste and Odor Improvement**: Activated carbon effectively enhances the taste and odor of harvested rainwater.

3. **Efficacy**: While activated carbon filters boast high efficacy for specific pollutants, they require regular replacement to maintain effectiveness.

   - **Replacement Schedule**: Follow manufacturer guidelines for replacement intervals for consistent performance.

   - **Cost Considerations**: The recurring cost of filter replacement should be factored into the overall system maintenance budget.

   - **Temperature Sensitivity**: The efficacy of activated carbon filters is influenced by water temperature, and you should consider this factor during system design.

UV Sterilization

1. **Mechanism**: UV sterilization harnesses ultraviolet light to disrupt the DNA of microorganisms, rendering them unable to

reproduce and guaranteeing a microbial-free water supply.

- **Microbial Disruption:** The targeted use of UV light effectively neutralizes bacteria, viruses, and other microorganisms.

- **UV-C Wavelength:** UV sterilization typically utilizes UV-C light, which is particularly effective in microbial inactivation.

2. **Optimal Applications:** UV sterilization shines in scenarios where microbial disinfection is the primary concern, offering a technology-driven solution for water safety.

- **Microbial Threat Mitigation:** Ideal for applications where the risk of microbial contamination is high.

- **Post-Filtration Disinfection:** UV sterilization is often employed as a post-filtration step to ensure the microbiological safety of harvested rainwater.

3. **Efficacy:** Highly effective for microbial disinfection and UV sterilization. However, it falls short when it comes to removing particulate matter or chemical contaminants.

- **Additional Filtration:** Combining UV sterilization with other filtration methods addresses the comprehensive removal of impurities.

- **Energy Consumption:** Consider the energy requirements associated with UV sterilization systems.

- **Installation Considerations:** Proper installation and regular maintenance are crucial for the sustained efficacy of UV sterilization systems.

Reverse Osmosis

1. **Mechanism:** Reverse osmosis employs a semipermeable membrane to remove ions, molecules, and larger particles from the water, offering a comprehensive solution to diverse contaminants.

- **Membrane Technology:** The semipermeable membrane acts as a molecular sieve, allowing only pure water molecules to pass through.

- **Pressure Differential:** Reverse osmosis relies on a pressure differential to drive the separation of water from impurities.

2. **Optimal Applications:** Reverse osmosis shines in applications where a wide range of contaminants, including minerals, chemicals, and microorganisms, need to be effectively removed.

- **Diverse Contaminant Removal:** Suitable for scenarios where water purity is paramount, addressing mineral content, chemicals, and microbial threats.

- **Residential Use:** Commonly employed in residential settings for producing purified drinking water.

3. **Efficacy:** Highly efficient in removing diverse contaminants, reverse osmosis systems have lower water production rates and involve wastewater generation.

- **Water Production Rate:** Users should be mindful of the potential impact on water availability, especially in areas with limited rainfall.

- **Wastewater Considerations:** Reverse osmosis systems generate wastewater, and proper disposal or reuse strategies should be in place.

- **Mineral Removal:** While effective in removing minerals, you may need to consider remineralization methods for the produced water.

The effectiveness of filtration techniques lies in their application. From the robust defense of mesh filters against larger particles to the precision of reverse osmosis in tackling a spectrum of impurities, each method contributes to water safety.

Harvesting rainwater for consumption or various domestic purposes demands a tailored approach. A well-designed filtration system, incorporating the strengths of different techniques, ensures the removal of specific contaminants and the overall purity and safety of the harvested rainwater.

Disinfection and Regular Maintenance

Filtration is the vanguard in harvesting clean rainwater, but the quest for purity doesn't end there. Post-storage treatments and regular maintenance emerge as unsung heroes, ensuring the sustained quality of harvested rainwater. In this section, you'll discover various disinfection methods, such as chlorination and ozonation. Additionally, you'll see the intricacies of regular maintenance practices that form the backbone of a resilient rainwater harvesting system.

Chlorination

1. **Mechanism:** Chlorination, a time-tested method, involves the introduction of chlorine or chlorine-based compounds into water. This chemical disinfection process is designed to annihilate microorganisms and safeguard water quality.

   - **Chlorine Types:** Chlorine is applied in various forms, including chlorine gas, liquid sodium hypochlorite, or solid calcium hypochlorite.

   - **Microbial Neutralization:** Chlorine disrupts the cellular structures of bacteria and viruses, rendering them inactive and preventing waterborne diseases.

2. **Optimal Applications:** Chlorination finds its sweet spot in scenarios where continuous disinfection of stored water is paramount.

   - **Tank Disinfection:** Applied to water stored in tanks, chlorination ensures a consistent level of microbial control.

   - **Public Water Systems:** Widely used in municipal water treatment, chlorination is a staple for ensuring potable water.

3. **Considerations:** While chlorination is a potent disinfection method, careful dosing is crucial to prevent over-chlorination.

   - **Dosing Control:** Precise control of chlorine dosage is essential to avoid health risks associated with excess chlorine in drinking water.

   - **Chemical Contaminant Limitations:** Chlorination may not effectively remove certain chemical contaminants, necessitating additional filtration measures.

   - **Residual Chlorine Management:** Managing residual chlorine levels is crucial to ensure water safety and prevent undesirable taste or odor.

Ozonation

1. **Mechanism:** Ozonation introduces ozone, a powerful oxidizing agent, into water. Ozone's oxidative prowess disinfects water by neutralizing contaminants and microorganisms.

- **Ozone Generation:** Ozone is generated on-site using specialized ozone generators, ensuring freshness and efficacy.

- **Pathogen Inactivation:** Ozone effectively inactivates bacteria, viruses, and some organic pollutants, contributing to comprehensive water safety.

2. **Optimal Applications:** Ozonation shines in applications where a broader spectrum of contaminants, including bacteria and viruses, must be addressed.

- **Microbial Disinfection:** Ozone serves as a robust defense against pathogenic microorganisms, ensuring waterborne disease prevention.

- **Organic Pollutant Removal:** Ozone's ability to break down organic pollutants enhances its efficacy in improving water quality.

3. **Considerations:** Implementing ozone systems requires careful calibration and attention to specific considerations to ensure optimal performance.

- **Calibration Precision:** Ozone systems demand precise calibration to achieve the desired disinfection efficacy without compromising water safety.

- **Residual Ozone Management:** Managing residual ozone levels is crucial, as excessive ozone in the water is harmful and impacts the water's taste.

- **Complexity of Installation:** Ozonation systems, while effective, are complex to install and may require professional assistance.

Regular Maintenance Practices

Ensuring the long-term safety of harvested rainwater involves the diligent integration of regular maintenance practices. These practices form a proactive shield against potential threats to water quality, guaranteeing a consistent supply of clean and safe rainwater.

Tank Inspection

- **Regularity:** Tanks should be routinely inspected for signs of wear, corrosion, or damage that could compromise water quality.

- **Seal Integrity:** Make sure that tank seals and joints are intact, preventing the infiltration of external contaminants.
- **Coating Integrity:** Check the integrity of any coatings on the tank interior, addressing any degradation promptly.

Cleaning Gutters and Screens

- **Frequency:** Regularly clean gutters and mesh filters to prevent the buildup of debris that compromises water quality.
- **Mesh Integrity:** Closely monitor the mesh integrity of filters, repairing or replacing damaged sections promptly.
- **Efficient Water Flow:** Unobstructed gutters and screens maintain efficient water flow, reducing the risk of contamination.

Filter Replacement

- **Adherence to Schedule:** Replace filters as recommended by the manufacturer to maintain their efficacy.
- **Filter Type Consideration:** Consider the specific type of filter and its lifespan, adjusting the replacement schedule accordingly.
- **Documentation:** Keep a record of filter replacements to facilitate a proactive and organized maintenance approach.

Flushing the System

- **Periodicity:** Periodically flush the system to remove stagnant water and potential contaminants.
- **System Efficiency:** Flushing enhances system efficiency by preventing the buildup of sediments and microbial growth.
- **Water Quality Assurance:** Regular flushing contributes to consistent water quality, especially in systems with infrequent use.

The essence of water purity lies in diligence and proactive care. Chlorination and ozonation stand as stalwart guardians, neutralizing threats at the chemical and microbial levels. Regular maintenance practices ensure the long-term integrity of the rainwater harvesting system.

The synergy of these approaches echoes the principles of sustainability and environmental stewardship. The journey doesn't end with the collection of rainwater. It extends into the treatment and maintenance, guaranteeing that every drop harvested remains a testament to the commitment to clean and sustainable water practices.

By addressing these aspects, you'll harness the full potential of rainwater harvesting, not only as a sustainable water source but also as a source of water purity. The commitment to water safety is not just a technical endeavor. It is a holistic approach that considers environmental factors, technology, and proactive maintenance.

Chapter 7: Beyond the Basics - Advanced Systems and Techniques

Water harvesting has evolved far beyond the rudimentary practices of the past. In this chapter, you'll learn about cutting-edge technologies, innovative materials, and integrated systems that define the forefront of rainwater harvesting. These advanced methods optimize water collection and storage, seamlessly integrating with other sustainable practices and offering a holistic approach to water management in various climates and terrains.

Innovative Materials and Designs

Rainwater harvesting, once a simple practice reliant on basic roofing materials, has entered a new era of innovation. Recent advancements in materials and designs are reshaping how people collect and utilize rainwater. In this section, you'll explore three groundbreaking innovations that are transforming rainwater harvesting.

Aerogels

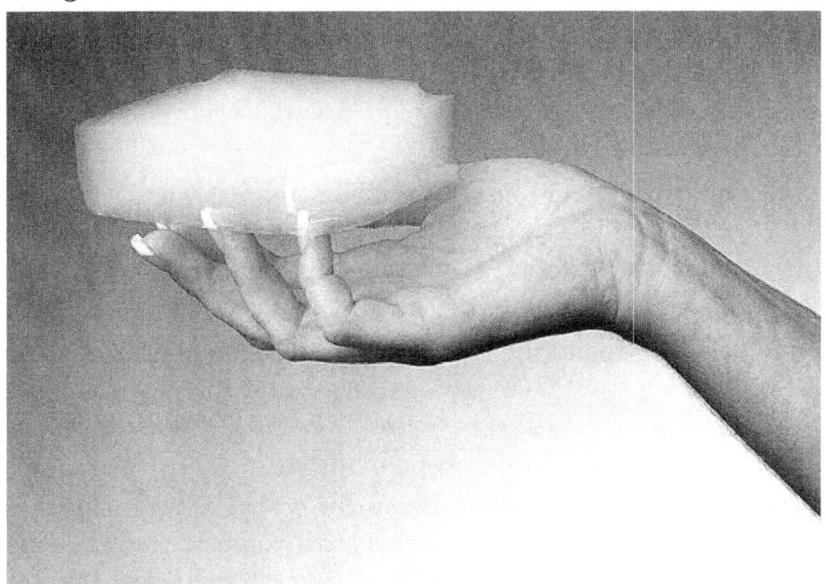

Aerogels are revolutionary materials with a porous and lightweight structure that enhances water collection efficiency to new heights.

Traditionally, the effectiveness of rainwater harvesting depended on the design of the collection surface. Enter aerogels, a revolutionary material with a porous and lightweight structure that enhances water collection efficiency to new heights.

- **Increased Surface Area for Enhanced Collection**: Aerogels, with their intricate structure, provide an expanded surface area, capturing more water droplets from the air than conventional materials.

- **Versatility for Retrofitting**: The lightweight nature of aerogels makes them an ideal choice for integration into existing roofing materials, allowing for easy retrofitting of structures.

- **Rapid Condensation for Maximum Yield**: The porous structure of aerogels facilitates rapid condensation, ensuring that even the smallest droplets are gathered efficiently, resulting in a substantial boost in overall water yield.

Smart Surfaces

Imagine surfaces that respond intelligently to environmental conditions, optimizing the entire rainwater harvesting process. Smart surfaces equipped with sensors and actuators make this vision a reality.

- **Real-Time Environmental Monitoring:** Smart surfaces are embedded with sensors that detect changes in temperature, humidity, and precipitation in real-time, allowing for adaptive responses.

- **Optimal Water Flow and Filtration:** Actuators on smart surfaces adjust properties like inclination and permeability, facilitating optimal water flow during heavy rainfall and enhanced filtration during lighter precipitation.

- **Remote Monitoring for Efficiency:** Integration with the Internet of Things (IoT) allows users to monitor the efficiency of their rainwater harvesting systems remotely, ensuring proactive maintenance and optimal water quality.

The marriage of aerogels and smart surfaces represents a leap forward in rainwater harvesting technology. It combines the efficiency of advanced materials and the adaptability of intelligent design, promising greater yields and sustainability.

Hydrophilic Coatings

One of the challenges in rainwater harvesting, especially in arid regions, is the low humidity that limits water collection. Hydrophilic coatings present a game-changing solution by giving surfaces the ability to attract and retain water molecules from the air.

- **Versatility for Various Materials:** Hydrophilic coatings can be applied to a variety of materials, including traditional roofing materials, offering a scalable solution for enhancing efficiency.

- **Promoting Water Adhesion:** The coatings create an environment that encourages water molecules to adhere, significantly increasing water capture even in low-humidity conditions.

- **Cost-Effective and Scalable:** The adaptability of hydrophilic coatings makes them a cost-effective and scalable solution for enhancing the efficiency of existing rainwater harvesting systems.

In addition to increasing water capture, hydrophilic coatings contribute to the prevention of water runoff. By promoting water adhesion, these coatings minimize wastage and maximize the potential for collection.

Architectural Marvels

In the world of rainwater harvesting, form is now meeting function through cutting-edge architectural designs. The integration of rainwater collection systems into buildings has given rise to structures that are both practical and aesthetically pleasing.

- **Self-Draining Roofs and Facades:** Innovative designs incorporate self-draining roofs that efficiently channel rainwater toward collection points, eliminating stagnant water and potential leaks.

- **Visual Harmony with Greenery:** Rooftops adorned with lush greenery contribute to a visually harmonious and eco-friendly environment where plants assist in water absorption and purification.

- **Interactive Building Elements:** Some architectural marvels go beyond passive rainwater collection, incorporating interactive elements that engage inhabitants in the water harvesting process.

Now that you have an idea of the technological marvels in the modern world, here's a closer look at each of them.

1. **Self-Draining Roofs and Facades:** Imagine a building with a roof that protects from the elements and actively contributes to water collection. Self-draining roofs are designed to channel rainwater efficiently toward collection points, eliminating stagnant water and potential leaks.

   - **Efficient Water Channeling:** These roofs are equipped with a slope and drainage system that ensures rainwater is directed towards collection points, preventing water from pooling and causing damage.

   - **Elimination of Stagnant Water:** By efficiently draining rainwater, self-draining roofs eliminate the risk of stagnant water, reducing the potential for leaks and structural damage.

   - **Enhanced Durability:** The design enhances water collection and contributes to the longevity of the roofing system, as pooling water is a common cause of deterioration.

2. **Visual Harmony with the Environment:** Architectural designs incorporate rainwater collection elements seamlessly into the

visual appeal of the structure. Rooftops adorned with lush greenery, where plants assist in water absorption and purification, contribute to a visually harmonious and eco-friendly environment.

- **Aesthetic Integration:** Rainwater collection elements are seamlessly integrated into the overall design of the building, enhancing its aesthetic appeal and contributing to a visually pleasing environment.

- **Green Rooftops for Biodiversity:** Rooftop gardens aid in rainwater absorption and create habitats for biodiversity, promoting ecological balance in urban environments.

- **Dual Functionality:** The integration of aesthetics with the rainwater collection provides dual functionality, making the building visually appealing and environmentally conscious.

3. **Interactive Building Elements:** Some architectural marvels go beyond passive rainwater collection. They incorporate interactive elements, such as transparent sections that showcase the flow of rainwater or kinetic features that respond to the volume of collected water. These designs serve a practical purpose and engage inhabitants in the water harvesting process.

- **Transparency for Education:** Transparent sections in building elements allow inhabitants to observe the flow of rainwater, promoting awareness and education about the importance of water conservation.

- **Kinetic Features for Engagement:** Kinetic elements, such as water features activated by collected rainwater, provide an engaging and interactive experience for inhabitants, fostering a sense of connection with the water harvesting process.

- **Educational and Recreational Value:** Interactive building elements contribute to the efficiency of rainwater harvesting and add educational and recreational value to the building, enhancing its overall significance in the community.

Advanced materials and designs are ushering in a new era for rainwater harvesting. Aerogels and smart surfaces offer unprecedented efficiency and adaptability. Hydrophilic coatings tackle the challenge of low-humidity environments. Architectural marvels redefine how you perceive the integration of water collection into your built environment.

As you embrace these innovations, you move toward a more sustainable and water-secure future.

Automation in Rainwater Harvesting

In the ever-evolving landscape of water management, automation has emerged as a beacon of efficiency, transforming rainwater harvesting into a smart and sustainable practice. In this section, you'll explore the three pillars of automation in rainwater harvesting. These advancements streamline the collection and storage process while paving the way for a more conscientious and resource-efficient approach to water utilization.

IoT Sensors

The integration of IoT into rainwater harvesting marks a paradigm shift in how people approach water resource management. IoT sensors, strategically placed on rooftops and within storage systems, serve as the eyes and ears of the entire system. They continuously gather real-time data on crucial parameters such as rainfall, water levels, and overall system health.

- **Real-Time Data for Informed Decision-Making:** These sensors empower users with a wealth of information at their fingertips. You'll access live data on the current rainfall intensity or the precise volume of water stored in your harvesting system from the comfort of your smartphone or computer. This real-time insight allows for informed decision-making, enabling you to optimize your harvesting systems based on current conditions.

- **Proactive Maintenance for System Longevity:** One of the significant advantages of IoT sensors is their ability to facilitate proactive maintenance. By monitoring the health of the system in real time, you identify potential issues before they escalate. Whether it's a clogged filter, a malfunctioning pump, or a structural concern, early detection allows for timely intervention, preserving the efficiency and longevity of the rainwater harvesting infrastructure.

- **Accessibility and Remote Monitoring:** The accessibility of real-time data is not limited by geographical constraints. You can remotely monitor your rainwater harvesting systems, making it especially beneficial for installations in remote or hard-to-reach locations. This capability enhances the overall efficiency of system management by enabling quick responses to changing conditions, regardless of physical proximity.

Automated Filtration Systems

Rainwater, while a valuable resource, is not immune to impurities. Automated filtration systems equipped with advanced technologies are revolutionizing the way you ensure the purity of harvested water.

- **Self-Cleaning Mechanisms for Uninterrupted Performance:** Traditional filtration systems often require manual intervention for cleaning and maintenance. Automated filtration systems, on the other hand, feature self-cleaning mechanisms that keep the filters free from contaminants. It guarantees a continuous supply of clean water and minimizes the need for frequent and labor-intensive upkeep.

- **Adapting to Varying Water Qualities:** Water quality varies based on factors such as climate, seasonal changes, and environmental influences. Automated filtration systems are designed to adapt dynamically to these variations. Whether the water source is experiencing a sudden influx of debris during heavy rainfall or a change in composition during dry spells, these systems adjust their filtration processes to maintain optimal performance.

- **Integration with Water Quality Sensors:** The automation of filtration systems can be further enhanced through integration with water quality sensors. These sensors detect specific impurities or contaminants in the harvested water. The filtration system then adjusts its processes in response to real-time data, ensuring that the water meets the desired quality standards. This level of precision in water purification enhances the overall reliability and safety of the harvested water.

Smart Distribution Networks

Automation doesn't end with collection and storage. It extends its transformative touch to the distribution phase. Smart distribution networks leverage automation to regulate the flow of water based on demand and availability.

- **Regulating Flow with Smart Pumps and Valves:** In traditional rainwater harvesting setups, water distribution often relies on manual adjustments or fixed schedules. Smart pumps and valves, powered by automation, bring a new level of precision to this process. These intelligent components regulate the flow of water based on real-time demand, ensuring a constant and reliable supply.

- **Minimizing Energy Consumption for Sustainability:** Energy efficiency is a cornerstone of sustainable water management. Smart distribution networks excel in this aspect by minimizing energy consumption. Pumps and valves operate precisely when needed, avoiding unnecessary energy expenditure. It contributes to environmental sustainability and translates into cost savings for you.

- **Predictive Analytics for Optimal Resource Allocation:** The integration of predictive analytics into smart distribution networks adds another layer of sophistication. By analyzing historical usage patterns and considering environmental factors, these systems predict future demand with remarkable accuracy. This predictive capability allows for proactive adjustments in water distribution, optimizing resource allocation and ensuring a consistent supply even during periods of high demand.

The marriage of IoT sensors, automated filtration systems, and smart distribution networks forms a trinity that propels rainwater harvesting into a realm of holistic sustainability. This synergy optimizes the efficiency of water collection and ensures the purity of the harvested water and its judicious distribution.

As you embrace the era of automation in rainwater harvesting, you pave the way for a future where water is managed with precision, sustainability, and a profound understanding of its vital role in your life. The journey towards water resilience has never been more promising.

Integrated Sustainable Systems

In the ever-evolving landscape of sustainable water management, the integration of various systems yields powerful and synergistic results. This section will shine a light on three interconnected approaches that exemplify the symbiosis between rainwater harvesting and other sustainable practices.

Greywater Recycling

Rainwater harvesting, when coupled with greywater recycling, establishes a holistic water management system that maximizes every precious drop. Greywater, originating from domestic activities like bathing and laundry, is a valuable resource often underutilized. The integration of greywater with rainwater harvesting creates a harmonious relationship where both water sources complement each other.

- **Tapping into Greywater's Potential:** While greywater is not suitable for drinking, its potential for non-potable uses is immense. Properly treated, greywater can be seamlessly combined with harvested rainwater for applications like irrigation. This dual approach significantly reduces reliance on external water sources for non-drinking purposes, a pivotal step towards sustainable water utilization.

- **Treatment and Purification Strategies:** Greywater treatment involves the removal of impurities and contaminants to meet the required quality standards for its intended use. Technologies such as filtration and biological treatment systems are employed to purify greywater. When merged with harvested rainwater, this treated blend becomes a versatile and eco-friendly water source for maintaining gardens, lawns, and other non-potable water needs.

- **Reducing Environmental Impact:** Beyond its immediate benefits, integrating greywater recycling with rainwater harvesting can potentially reduce the environmental impact associated with water consumption. By minimizing reliance on conventional water sources, this combined system contributes to the conservation of freshwater ecosystems and mitigates the strain on municipal water supplies.

- **Accessibility and Remote Monitoring:** The accessibility to real-time data is not limited by geographical constraints. You can remotely monitor your rainwater harvesting systems. This capability enhances the overall efficiency of system management by enabling quick responses to changing conditions, regardless of physical proximity.

Permaculture Designs

Permaculture, deeply rooted in principles of sustainability and harmonious coexistence with the environment, offers a natural companion to rainwater harvesting. Integrating these practices allows for the creation of self-sustaining ecosystems that promote soil health, biodiversity, and overall environmental well-being.

- **The Role of Rainwater in Permaculture:** Rainwater, as a fundamental component of permaculture designs, is harnessed to nurture the landscape. Various techniques, such as swales and contour bunds, are employed to capture and direct rainwater to where it is needed most. It guarantees efficient water utilization and prevents soil erosion while aiding in the cultivation of diverse plant species.

- **Regenerative Agriculture Practices:** The integration of rainwater harvesting and permaculture extends to regenerative agriculture practices. You'll foster healthier soil conditions by capturing rainwater and implementing permaculture techniques. This, in turn, enhances crop resilience, reduces the need for synthetic fertilizers, and contributes to the restoration of degraded land.

- **Biodiversity Promotion:** Permaculture emphasizes the importance of biodiversity in agricultural and natural systems. A more resilient and diverse ecosystem can be created by incorporating rainwater harvesting into permaculture designs. The stored rainwater provides a lifeline during dry spells, fostering the survival of various plant and animal species.

- **Community Engagement and Education:** The integration of permaculture and rainwater harvesting is not only about cultivating sustainable ecosystems but also about community engagement and education. By sharing knowledge and practices, communities can collectively work towards a more sustainable and regenerative way of living.

Aquaponics and Rainwater

The marriage of aquaponics and rainwater harvesting presents an innovative approach to sustainable agriculture. In aquaponics, fish and plants form a symbiotic relationship where fish waste provides essential nutrients for plants. The plants, in turn, purify the water for the fish. When seamlessly integrated with rainwater harvesting, this system becomes a closed-loop, resource-efficient powerhouse.

- **Rainwater as the Lifeblood of Aquaponics:** Rainwater, collected and stored, serves as the nutrient-rich lifeblood of aquaponics systems. You'll reduce your dependence on external water sources by utilizing rainwater in aquaponics.

- **Resource Efficiency and Closed-Loop Systems:** The integration of aquaponics and rainwater harvesting embodies the essence of resource efficiency. Fish waste, a natural byproduct of aquaponics, becomes a fertilizer for plants. As the plants absorb these nutrients, they contribute to the purification of the water, creating a closed-loop system that minimizes waste and maximizes efficiency.

- **A Blueprint for Urban Agriculture:** The compact nature of aquaponics makes it particularly suitable for urban agriculture. By incorporating rainwater harvesting, you create sustainable and self-sufficient growing systems. This approach reduces the environmental footprint associated with conventional agriculture and provides a blueprint for cultivating fresh produce in limited urban spaces.

- **Educational Opportunities and Food Security:** The integration of aquaponics and rainwater harvesting also offers educational opportunities and contributes to food security. By promoting this sustainable farming method, you learn about the interdependence of ecosystems and gain valuable skills in urban agriculture, fostering a sense of food autonomy.

Smart Landscaping

Among the many facets of integrated sustainable systems, smart landscaping emerges as a pivotal player, seamlessly blending aesthetics with purpose. By incorporating rainwater harvesting into landscaping designs, you create outdoor spaces that captivate the eye and contribute to environmental sustainability.

- **Rain Gardens and Sustainable Landscapes:** Rain gardens, strategically designed to capture and manage stormwater runoff, exemplify the union of landscaping and rainwater harvesting. These gardens are engineered to absorb rainwater, preventing soil erosion and minimizing the flow of contaminants into water bodies. By integrating rain gardens into landscaping plans, you'll transform your outdoor spaces into dynamic elements of a sustainable water management system.

- **Edible Landscapes and Urban Agriculture:** The concept of edible landscapes takes landscaping a step further, integrating ornamental plants with edible ones. By combining rainwater harvesting with edible landscaping, you'll cultivate fruits, vegetables, and herbs using collected rainwater. This approach provides a sustainable source of fresh produce and reduces the carbon footprint associated with transporting food from distant locations.

- **Biodiverse and Water-Efficient Plant Selection:** Smart landscaping extends to the careful selection of plants that thrive in specific climates with minimal water requirements. By choosing native and drought-resistant plants, you'll contribute to water efficiency and biodiversity. This intentional selection aligns with the principles of permaculture, promoting a harmonious relationship between human-managed spaces and the natural environment.

As you explore the integration of rainwater harvesting with greywater recycling, permaculture designs, aquaponics, and smart landscaping, a common thread emerges. It's the pursuit of harmony with the environment. These integrated sustainable systems optimize water utilization and foster regenerative practices that contribute to the planet's well-being.

Toward a Sustainable Future

The journey toward a sustainable future involves embracing diverse practices that work in tandem to conserve resources and nurture ecosystems. The integration of rainwater harvesting with greywater recycling, permaculture designs, and aquaponics exemplifies this harmonious approach. By adopting these integrated systems, individuals, communities, and agricultural practitioners can become stewards of a more sustainable and resilient world.

The future of water management lies in synergy. Innovative materials and designs enhance the efficiency of collection structures. Automation streamlines the entire process, and integration with other sustainable practices creates holistic solutions. The adaptability of these advancements makes them applicable in various climates and terrains, empowering you to harness the power of rainwater most efficiently and sustainably.

Chapter 8: Nature's Bounty – Uses for Your Harvest

Rainwater harvesting opens the door to a wealth of possibilities, turning every droplet into a valuable resource. In this chapter, you'll explore the myriad applications of harvested rainwater, underscoring its versatility and the potential to redefine how you meet your water needs. From everyday household uses to agricultural benefits and the possibilities of potable water, you'll uncover the vast potential nature's bounty holds.

Distinguishing Potable and Non-Potable Uses

Before exploring the diverse applications of harvested rainwater, it's crucial to distinguish between potable and non-potable uses. Potable water is suitable for drinking, while non-potable water is used for other purposes, such as irrigation or cleaning. The water quality required for these applications varies, with potable water demanding the highest standards to guarantee human health.

To transform harvested rainwater into a safe drinking source, stringent quality parameters must be met. It involves thorough filtration, disinfection, and monitoring to eliminate contaminants and pathogens. Compliance with regulatory standards ensures that the water is potable and meets the highest safety requirements.

Household Applications

In sustainable living, harvested rainwater emerges as a versatile and invaluable resource, revolutionizing everyday household applications. From laundry to cleaning and gardening, rainwater's soft and chemical-free composition transforms mundane chores into eco-friendly and economically sustainable practices. Here's how rainwater can enhance the fabric of your daily life.

Laundry

Laundry day takes on a whole new dimension with the introduction of harvested rainwater.
https://pixabay.com/vectors/washhouse-laundry-house-room-294621/

Laundry day takes on a whole new dimension with the introduction of harvested rainwater. Unlike hard water from conventional sources, rainwater is naturally soft, devoid of harsh minerals that compromise the effectiveness of detergents and take a toll on fabrics. This inherent softness creates a gentle, nurturing environment for clothing, ensuring a softer touch and prolonging the life of each garment.

- **Enhancing Detergent Efficiency:** The soft nature of rainwater enhances the efficiency of detergents, allowing them to lather more effectively and penetrate fabric fibers with ease. Even in areas with hard tap water, where detergents struggle to reach their full potential, rainwater provides a solution that maximizes cleaning power while minimizing the use of chemical additives.

- **Prolonging Clothing Longevity:** Hard water, laden with minerals like calcium and magnesium, contributes to the wear and tear of clothing over time. The abrasive effects of these minerals on fabric fibers lead to fading, reduced softness, and a shorter lifespan for garments. By opting for rainwater in laundry, you'll invest in the longevity of your clothing, reducing the frequency of replacements and contributing to a more sustainable wardrobe.

- **Reducing Environmental Impact:** Beyond the benefits to fabrics, using rainwater for laundry contributes to environmental sustainability. Traditional water sources often require extensive treatment processes and transport, consuming energy and contributing to carbon emissions. In contrast, rainwater harvesting for laundry significantly reduces the environmental impact associated with water use, aligning with eco-conscious living practices.

Cleaning

Household cleaning takes on a new dimension when nature's purifier, rainwater, becomes the cleaning agent of choice. Its soft and chemical-free composition makes it ideal for a variety of cleaning purposes, from surfaces and windows to delicate items that require a tender touch.

- **Ideal for Surfaces and Windows:** The softness of rainwater makes it particularly effective for cleaning surfaces and windows. Without the minerals present in hard water that leave streaks and residue, rainwater leads to a crystal-clear finish. Whether it's wiping down countertops, glass tables, or mirrors, rainwater leaves surfaces spotless, all while being gentle on the materials being cleaned.

- **Preserving Delicate Items:** Certain delicate items, such as intricate glassware, fragile decorations, or heirloom pieces, benefit from the gentle touch of rainwater. The absence of harsh chemicals ensures that delicate surfaces remain unharmed during the cleaning process. It preserves the integrity of these items and reflects a commitment to sustainable cleaning practices.

- **Reducing Reliance on Tap Water:** Harnessing rainwater for cleaning needs reduces reliance on tap water, contributing to both environmental and economic sustainability. The energy-intensive processes involved in treating and distributing tap water are minimized, resulting in a reduced carbon footprint. This shift towards rainwater for cleaning aligns with the broader movement toward responsible water use.

Gardening

Perhaps one of the most rewarding applications of harvested rainwater is in the garden. Rainwater, free from chlorine and other additives commonly found in tap water, nurtures plants with the pure essence of hydration. The controlled pH levels in rainwater make it a perfect match for various plant species, promoting healthy growth and vibrant blooms.

- **Chlorine-Free Hydration**: Many municipal water supplies are treated with chlorine to eliminate bacteria and pathogens. While this is essential for human consumption, plants do not share the same affinity for chlorine. Rainwater, being free from this chemical additive, provides plants with a chlorine-free source of hydration, promoting optimal growth and development.

- **Controlled pH Levels for Plant Health**: Rainwater typically has a slightly acidic pH, which is beneficial for certain plants that thrive in acidic soil conditions. This controlled pH level ensures that rainwater complements the specific needs of a variety of plant species, fostering an environment where they flourish. This natural compatibility makes rainwater ideal for watering gardens and potted plants.

- **Promoting Water Efficiency in Gardening**: Water efficiency in gardening is a key consideration for sustainable practices. Rainwater harvesting directly addresses this concern by providing a local and on-site water source for plants. By utilizing rainwater for gardening, you'll reduce the demand for municipal water supplies, contributing to water conservation efforts and promoting a more resilient and eco-friendly landscape.

As you navigate the waters of sustainable living, the incorporation of harvested rainwater into household applications reveals a gentle yet transformative touch. From the softer embrace it offers to fabrics to its role as nature's purifier in cleaning and the growth it conducts in the garden, rainwater emerges as a precious resource that goes beyond mere functionality.

Agricultural and Landscaping Benefits

In the vast tapestry of sustainable water management, the agricultural and landscaping sectors stand as canvas and garden alike, awaiting the transformative touch of harvested rainwater. This precious resource, harvested and stored during moments of abundance, is a lifeline for crops and a nurturing elixir for landscapes. This section scours the agricultural realm, where rainwater becomes a sustainable alternative for farming, and explores how landscaping transforms into natural masterpieces under its gentle care.

Farming

In the agricultural expanse, water is the lifeline that sustains crops, ensuring their growth, health, and productivity. Traditional sources, however, often come with challenges of fluctuating water availability, reliance on distant reservoirs, and the need for energy-intensive irrigation systems. This is where rainwater harvesting steps in as a sustainable alternative, offering a locally sourced and environmentally friendly solution.

- **Safeguarding Crops from Drought Stress:** The unpredictable nature of weather patterns, including periods of drought, poses a significant threat to agricultural productivity. Rainwater harvesting provides a buffer against these challenges by enabling farmers to store and utilize rainwater during dry spells. This stored bounty becomes a lifeline for crops, offering a consistent water supply that safeguards them from the stress of water scarcity.

- **Enhancing Soil Health with Pure Rainwater:** Beyond just providing hydration, rainwater contributes to the overall health of agricultural lands. Unlike traditional water sources treated with chlorine and other additives, rainwater is pure and free from chemical interventions. This purity extends to the soil, enhancing its health and fertility. The absence of chemical residues ensures that the soil becomes a thriving ecosystem where beneficial microorganisms flourish, supporting the vitality of crops.

- **Promoting Sustainable Agriculture Practices:** Rainwater harvesting aligns seamlessly with the principles of sustainable agriculture. By relying on a locally sourced and naturally replenished water supply, you'll reduce your dependence on external water sources. It minimizes the environmental impact associated with long-distance water transport and promotes a more self-sufficient and resilient agricultural system.

Landscaping

Landscaping, whether in residential gardens or expansive public spaces, transforms into a canvas of natural masterpieces when nourished by harvested rainwater. The purity and softness of rainwater offer a gentle touch that promotes the health and vibrancy of plants. Free from the harsh minerals and additives found in conventional water sources, rainwater becomes a nurturing elixir for the green elements of your surroundings.

- **Reducing the Need for Chemical Interventions:** Conventional water sources often contain minerals that, over time, accumulate in the soil and affect plant health. Harsh elements like calcium and magnesium present in hard water necessitate chemical interventions to counter their impact. Rainwater, with its soft and mineral-free composition, eliminates the need for such interventions, allowing plants to thrive naturally.

- **Efficient Irrigation for Thriving Landscapes:** The controlled distribution of rainwater through efficient irrigation systems is a key element in crafting natural masterpieces in landscaping. Rainwater harvesting systems, equipped with intelligent irrigation mechanisms, ensure that landscapes receive just the right amount of water. This precision promotes water efficiency, preventing overwatering and minimizing runoff. It contributes to an eco-friendly and sustainable landscaping approach.

- **Enhancing Biodiversity and Ecological Balance:** The soft touch of rainwater extends beyond individual plants to the broader ecosystem within landscapes. By reducing the reliance on conventional water sources, which are treated with chemicals for purification, rainwater contributes to the preservation of biodiversity. Beneficial insects, microorganisms, and other components of the ecological balance within the landscape thrive in an environment free from the adverse effects of chemical-laden water.

The synergy between rainwater harvesting in agriculture and landscaping creates integrated solutions that enhance water efficiency on multiple fronts. The same harvested rainwater that nourishes crops is directed to landscaping elements, fostering a harmonious approach to water use. This integration aligns with the principles of permaculture, where diverse elements coexist and contribute to a self-sustaining ecosystem.

The adoption of rainwater harvesting in these sectors is not merely a practical choice but a commitment to a future where water is valued. It's where ecosystems flourish, and landscapes become vibrant expressions of ecological harmony. By embracing rainwater as a cornerstone of agricultural and landscaping practices, you'll cultivate a sustainable legacy that respects the delicate balance of nature and ensures a resilient future for generations to come.

Potable Possibilities

The delicate alchemy of transforming harvested rainwater into potable elixir is a journey guided by stringent quality assurance measures. Filtration systems, ultraviolet treatments, and unwavering compliance with regulatory standards become the sentinels. These ensure that the pure essence of rainwater emerges as a thirst-quencher and as a paragon of safety.

Quality Assurance for Drinking Water

The metamorphosis of rainwater into potable water demands precision and dedication to quality. From the serene descent of raindrops to the moment it cascades into a drinking glass, every step is full of responsibility. The commitment to quality assurance transforms rain's embrace into a source of life that not only hydrates but nourishes without compromise.

- **Filtration Systems:** Harvested rainwater, while pristine in its origin, carries microorganisms and chemicals that pose potential risks to human health. Robust filtration systems are the first line of defense, serving as guardians against contaminants. Mesh filters, adept at trapping larger particles, and carbon filters, capable of removing impurities and odors, form a formidable alliance. These filters work in harmony to ensure that the water undergoes a transformative cleansing, emerging free from visible and invisible intruders.

- **Ultraviolet Treatments:** Beyond the physical aspects of filtration lies the illuminating touch of ultraviolet (UV) treatments. UV disinfection becomes a crucial step in the purification process. It targets microorganisms that persist despite the initial filtration. The power of UV light disrupts the DNA of bacteria, viruses, and other pathogens, rendering them incapable of causing harm. Once touched by the sun's rays, this final purification ensures that the water emerges as a microbiologically safe liquid.

- **Regulatory Standards:** In the vast expanse of potable possibilities, adherence to regulatory standards becomes the North Star, guiding the entire process. Local health regulations outline the benchmarks for water quality. They set the stage for a process where safety is non-negotiable. Understanding and incorporating these standards into the purification process guarantees the final product quenches thirst without compromising health.

Microbial and Chemical Safety Measures

The purity that makes rainwater a source of wonder also introduces challenges. Harvested rainwater carries microorganisms and chemicals that, if left unchecked, compromise its safety for human consumption. Recognizing and addressing these challenges becomes imperative in crafting potable possibilities from rain's embrace.

- **Robust Filtration Systems:** The journey toward potability begins with robust filtration systems that serve as the first line of defense. Mesh filters, with their intricate weave, capture larger particles and debris, preventing them from tainting the water. Carbon filters, with their porous structure, adsorb impurities, odors, and some chemicals, further enhancing the clarity and purity of the water. Together, these filtration systems form a formidable barrier against contaminants.

- **UV Disinfection:** The microbial safety measures reach their pinnacle with UV disinfection—a process that harnesses the power of light to neutralize microorganisms. The short-wavelength UV-C light damages the DNA of bacteria, viruses, and other pathogens, rendering them unable to reproduce or cause infection. This layer of protection ensures that the final product is not just visually clear but microbiologically safe.

- **Regular Testing:** The journey from raindrop to drinking glass is an ongoing commitment to safety. Regular testing for microbial and chemical parameters becomes a continuous assurance of the water's quality. Rigorous testing protocols, including checks for bacteria, viruses, and chemical compositions, guarantee that the potable possibilities of harvested rainwater remain steadfast in their commitment to health and safety.

As you savor the potable possibilities of harvested rainwater, let it be a reminder of the delicate balance between nature's bounty and your responsibility to safeguard health. In each sip, you taste not just the freshness of rain but the culmination of a journey. From the cloud-kissed skies to the vigilant filtration systems, from the dance of UV light to the commitment to regulatory standards, it's a journey where the possibilities of rainwater extend beyond nourishment. This liquid gold becomes a celebration of purity, a testament to human ingenuity, and a harmonious dance with the essence of life itself.

Environmental and Economic Benefits

In the delicate balance between nature's bounty and human needs, the harvesting of rainwater emerges as a transformative practice, ushering in a wave of environmental and economic benefits. At the heart of this practice lies the profound impact of reducing strain on municipal water supplies, mitigating stormwater runoff, and unveiling a water-wise approach that extends financial savings to the water-wise wallet.

- **Conserving Treated Water for Essential Uses:** One of the primary benefits of harvested rainwater lies in its potential to alleviate the strain on municipal water supplies. By meeting non-potable needs with naturally sourced water, you'll contribute significantly to the conservation of treated water for essential uses. This prudent conservation approach ensures that the limited and precious resource of treated water is reserved for purposes that demand the highest quality.

- **Easing the Burden on Water Treatment Facilities:** As rainwater becomes a readily available source for activities like gardening, cleaning, and irrigation, the burden on water treatment facilities eases. These facilities are designed to purify water to meet rigorous drinking standards. By diverting non-potable demands to harvested rainwater, you play a proactive role in preserving the integrity of treated water. It optimizes the efficiency of treatment processes and extends the lifespan of water infrastructure.

- **Reducing Energy Demands for Water Transportation**: The journey of water from treatment facilities to homes involves significant energy consumption, especially when transported over long distances. Harvested rainwater, sourced locally, disrupts this energy-intensive cycle. Utilizing water where it falls reduces the need for extensive transportation, contributing to a more sustainable and energy-efficient water supply system.

Proactive Management of Excess Water

Beyond its role in reducing strain on municipal water supplies, rainwater harvesting plays a crucial role in mitigating stormwater runoff. Stormwater runoff, often a culprit in urban flooding, results from rainfall that exceeds the absorption capacity of soil and surfaces. Harvesting rainwater at its source transforms homeowners and communities into proactive managers of excess water. You'll prevent soil erosion and minimize the flow of pollutants into rivers and streams.

- **Preserving Soil Health and Water Purity:** As rainwater is captured and directed for various uses, it permeates into the soil, replenishing aquifers and preserving soil health. It mitigates the risk of soil erosion and prevents the rapid flow of rainwater over impermeable surfaces, reducing the chances of water pollution. By becoming a steward of rain's descent, you'll embrace a holistic approach to water management that safeguards both the environment and its inhabitants.

- **Shrinking Water Bills through Smart Choices:** The economic benefits of harvesting rainwater extend beyond the environmental realm, offering tangible savings to households. As reliance on municipal water for non-potable needs diminishes, so do water bills. The initial investment in a rainwater harvesting system becomes a wise and enduring economic choice, creating a sustainable and cost-effective water supply.

- **Long-Term Economic Gain:** While there is an initial investment in installing rainwater harvesting systems, the long-term economic gain is substantial. The reduction in water bills, coupled with the potential for local government incentives or rebates for sustainable practices, transforms rainwater harvesting into a financially savvy choice. As you witness your water-wise wallet grow, the economic viability of rainwater harvesting becomes increasingly apparent.

Maximum Yield

When you get into sustainable water management, harvesting maximum yield from rainwater presents itself as a transformative and empowering practice. At its core, this journey encompasses the implementation of efficient rainwater harvesting systems, the optimization of storage capacities, and the embrace of water-wise practices. The union of environmental consciousness with practical application fosters individual empowerment and cultivates a community of stewards dedicated to making the most of every precious drop.

- **Designing for Maximum Capture**: At the heart of harvesting maximum yield is the thoughtful design and implementation of efficient rainwater harvesting systems. The journey begins with capturing rainwater at its source, be it on rooftops, surfaces, or catchment areas. Thoughtfully designed systems, equipped with advanced technologies and materials, ensure every raindrop is harnessed with precision.

- **Strategic Storage Solutions:** Optimizing harvested yield involves strategic storage solutions that align with the natural rhythms of rainfall. Robust storage capacities, whether in above-ground tanks, cisterns, or below-ground reservoirs, become the reservoirs of abundance. By maximizing storage, you store surplus rainwater for periods of scarcity, guaranteeing a consistent and reliable water supply throughout the year.

- **Smart Distribution Networks:** Equally important is the establishment of smart distribution networks within the storage system. Intelligent pumps, valves, and distribution mechanisms ensure that the stored rainwater is distributed efficiently, addressing the specific needs of different areas, whether for irrigation, gardening, or non-potable household uses. This strategic distribution optimizes the utility of harvested rainwater, maximizing its impact across various facets of daily life.

- **Landscape Design and Irrigation Efficiency:** Harvesting maximum yield extends beyond the technicalities of systems and storage. It embraces water-wise practices that cultivate conscious consumption. Thoughtful landscape design, incorporating native and drought-resistant plants, minimizes water demands. Efficient irrigation systems, such as drip irrigation or rain garden techniques, make it so that every drop is utilized properly, promoting a harmonious balance between nature and human needs.

- **Indoor Conservation Measures**: Water-wise practices also find their place indoors, where conscious consumption becomes a daily commitment. Installing low-flow fixtures, fixing leaks promptly, and embracing water-efficient appliances contribute to the overall goal of maximizing the utility of harvested rainwater. These measures amplify the impact of rainwater harvesting on reducing dependence on conventional water sources.

- **Workshops and Seminars**: Obtaining maximum yield is not a solitary endeavor. It thrives on educational initiatives and community engagement. Workshops and seminars become platforms for sharing insights into the benefits and applications of harvested rainwater. These provide you with the knowledge and tools needed to become an active participant in the journey toward sustainability.

- **Community Outreach Programs:** The ripple effect of change is amplified through community outreach programs. These initiatives foster a sense of community responsibility, encouraging individuals to become catalysts for positive change within their neighborhoods. By collectively embracing the principles of rainwater harvesting, communities transform into stewards of their water resources, nurturing a shared commitment to environmental sustainability.

Water-wise practices in outdoor landscaping and indoor consumption become the threads that weave conscious living. Yet, the journey is incomplete without the communal spirit fostered by educational initiatives and community engagement. As workshops and community outreach programs empower individuals with the knowledge to become stewards of their water resources, the collective impact becomes a force for positive change.

In concluding this exploration of nature's bounty and the uses for your harvest, the overarching theme is one of harmony and sustainability. From everyday household applications to agricultural benefits and the possibilities of potable water, harvested rainwater is a versatile and valuable resource. As you nurture nature's gift drop by drop, you take a step closer to a more sustainable, resilient, and water-wise future.

Chapter 9: Potable Rainwater: Making Your Harvest Drinkable

In the pursuit of self-sufficiency and sustainable living, transforming harvested rainwater into safe, potable water stands as a pinnacle of achievement. This chapter explores the intricate processes and necessary precautions required to make your rainwater harvest usable and drinkable. From understanding local water quality to employing advanced filtration methods and disinfection techniques, the journey toward potable rainwater is an exploration of both science and practicality.

In the pursuit of self-sufficiency and sustainable living, transforming harvested rainwater into safe, potable water stands as a pinnacle of achievement.

Understanding Local Water Quality

Every region, with its unique blend of environmental influences, industrial activities, and human settlements, has distinct challenges and characteristics that shape its water quality. This section explores the crucial significance of understanding these dynamics, emphasizing the awareness of potential contaminants, both natural and anthropogenic, as the foundational step toward ensuring the safety of the final drinking product.

Local Water Quality Dynamics

Nature's influence, industrial activities, and human settlements collectively shape the dynamic tapestry of local water quality. Understanding these intricate dynamics lays the groundwork for anticipating potential contaminants and tailoring purification strategies to the unique characteristics of each region.

- **Environmental Influences**: Nature, in all its diversity, plays a pivotal role in shaping the quality of local water sources. Environmental factors such as soil composition, topography, and vegetation contribute to the mineral content and overall characteristics of rainwater. For instance, water flowing through rocky terrain might carry higher mineral concentrations, impacting taste and safety. Understanding these natural influences provides a baseline for anticipating potential challenges in the purification process.

- **Industrial Activities**: Human activities, especially industrial processes, introduce a spectrum of substances into the local water supply. Runoff from industrial areas carries pollutants such as heavy metals, chemicals, and toxins. Awareness of nearby industrial activities is crucial in identifying potential contaminants that could seep into rainwater. This insight directs the selection of appropriate filtration and purification methods to address specific industrial pollutants.

- **Human Settlements**: Urban and rural settlements leave their imprints on local water quality. Urban areas introduce pollutants like pesticides, herbicides, and contaminants from vehicle emissions. In contrast, rural areas see agricultural runoff carrying fertilizers and pesticides into water sources. Understanding the footprint of human settlements enables a tailored approach to water purification, addressing the unique challenges posed by each environment.

Testing Protocols and Frequency

Testing is the guardian of the promise of safe drinking water. This section explores the essential parameters, from microbial content to chemical composition, and emphasizes the importance of a vigilant testing regime. Adhering to local regulations, adapting to seasonal variations, and monitoring changes in the surrounding environment are integral components of this ongoing commitment. A comprehensive testing regime should cover a spectrum of parameters critical to water quality:

- **Microbial Content:** Testing for bacteria, viruses, and other microorganisms assesses the microbial safety of the water. Coliform bacteria, for example, serve as indicators of fecal contamination and potential pathogenic risks.
- **Chemical Composition:** Analyzing the chemical makeup detects substances such as heavy metals, pesticides, and industrial pollutants. This step is vital in addressing both natural and anthropogenic contaminants.
- **Overall Water Quality:** Parameters like pH, turbidity, and dissolved oxygen contribute to the overall quality and usability of the water. Maintaining these within acceptable ranges ensures a safe and pleasant drinking experience.

Frequency of Testing

The frequency of testing is a dynamic aspect that adapts to local conditions and regulations:

- **Local Regulations:** Adherence to local regulations is paramount. Some regions have specific guidelines dictating the testing frequency for different parameters. Understanding and complying with these regulations establish a legal framework for ensuring water safety.
- **Seasonal Variations:** Seasonal changes influence water quality too. Increased agricultural activities during planting seasons or industrial processes in certain weather conditions elevate contamination risks. Adjusting the testing frequency to align with these variations ensures a proactive approach to potential

challenges.

- **Changes in Surrounding Environment:** Environmental shifts, such as nearby construction, changes in land use, or new industrial developments, impact water quality. Regular testing, especially during periods of environmental change, is an early warning system, allowing for prompt adjustments to purification methods.

A vigilant testing regime is a guardian that upholds the promise of safe drinking water. It is not a one-time affair but an ongoing commitment to monitor and adapt to the dynamic nature of local water quality. Regular testing serves as a proactive measure, allowing for timely adjustments in purification methods and ensuring the sustained potability of harvested rainwater.

Advanced Filtration Methods

Efficient filtration is the linchpin in the ambitious journey to transform rainwater into a potable resource. As you delve into the intricacies of advanced filtration methods, you'll encounter a world where the understanding of the size of microns becomes paramount. It's time to explore the science behind microns and their role in bacteria removal. You'll also uncover the pivotal contributions of activated carbon filters and reverse osmosis systems in ensuring a comprehensive and purified drinking experience.

Micron Sizes and Bacteria Removal

Understanding the size of microns is pivotal in designing filtration systems that act as the first line of defense against bacteria. Microfiltration and ultrafiltration systems, with their distinct pore sizes, lay the groundwork for thorough bacteria removal, ensuring the journey toward potable rainwater starts with precision.

- **The Microscopic World:** At the heart of advanced filtration is the microscopic realm of microorganisms, particularly bacteria. These tiny living beings, with sizes ranging from 0.2 to 5 microns, pose a significant challenge in the quest for potable rainwater. To effectively remove these threats, filtration systems must be strategically designed to capture particles within this size range.

- **Microfiltration:** Microfiltration systems, characterized by their relatively larger pore sizes compared to other advanced methods, serve as the initial defense against microorganisms. These filters typically have pores ranging from 0.1 to 10 microns, making them adept at trapping larger particles like bacteria. However, their efficacy varies, and additional filtration methods are often needed to ensure thorough removal.

- **Ultrafiltration**: Taking filtration precision to the next level, ultrafiltration systems boast smaller pore sizes, typically ranging from 0.002 to 0.1 microns. That enables them to capture even the smallest bacteria, providing a more comprehensive solution for microbial removal. Ultrafiltration, with its ability to target particles at the sub-micron level, lays the groundwork for achieving the stringent standards required for safe drinking water.

Activated Carbon Filters

Beyond microbial challenges, rainwater carries an array of chemical contaminants. Activated carbon filters step into the spotlight with their porous prowess, excelling at absorbing chemicals like chlorine, pesticides, and organic compounds. This section explores the magic of adsorption, the versatility of activated carbon in addressing various contaminants, and how this dual-action purification elevates the quality of harvested rainwater.

- **The Porous Powerhouse:** Activated carbon filters emerge as the unsung heroes in the battle against chemical impurities. Their porous structure, created through a process that activates carbon with steam or chemicals, provides an expansive surface area for adsorption.

- **Adsorption Magic:** Activated carbon's adsorption capacity is a game-changer in the purification process. As water passes through the filter, chemical contaminants adhere to the carbon surface, effectively removing them from the water. This dual-action purification, addressing both microbial and chemical impurities, elevates the quality of harvested rainwater to meet the high standards required for safe consumption.

- **Versatility in Application:** Activated carbon filters prove versatile in addressing a wide spectrum of chemical contaminants, including:

  a. **Chlorine:** Commonly used in water treatment but undesirable in drinking water due to taste and potential health concerns.

b. **Pesticides and Herbicides:** Agricultural runoff can introduce these chemicals into rainwater, posing risks to human health.

c. **Organic Compounds:** Various pollutants from industrial activities find their way into rainwater, necessitating effective removal for safety.

Reverse Osmosis Systems

In the pursuit of a comprehensive purification approach, reverse osmosis (RO) systems emerge as key players. Operating at the molecular level, RO utilizes a semi-permeable membrane to filter out impurities, from bacteria to dissolved salts and minerals. Here's how RO works:

1. **Semi-Permeable Membrane:** The heart of the RO system, the semi-permeable membrane, allows water molecules to pass through while blocking larger contaminants.

2. **Pressure Application:** Applying pressure to the water forces it through the membrane, separating impurities and pollutants.

3. **Reject Water Disposal:** The concentrated contaminants are then flushed away as rejected water, leaving behind purified water ready for consumption.

Understanding micron sizes is the compass guiding you through the microscopic world of bacteria removal, while activated carbon filters showcase their prowess in adsorbing chemical contaminants, ensuring a dual-action purification. Reverse osmosis systems provide a comprehensive purification approach that transcends bacterial threats to address molecular-level impurities. The combined efforts of these advanced filtration methods elevate harvested rainwater to a standard of potable purity, marking a triumph in the quest for sustainable and safe drinking water.

Disinfection Techniques

In the relentless pursuit of transforming harvested rainwater into potable water, the spotlight shifts to disinfection techniques. It's a critical phase where microbial safety takes center stage. From harnessing the power of light to embracing age-old practices, these techniques stand as guardians, ensuring the journey from raindrop to drinking glass is free from microbial threats.

UV Purification

In the radiant realm of modern disinfection, UV purification is a powerful and effective method. As you unveil the science behind harnessing the power of light, specifically UV-C, you'll witness a process that damages the DNA of microorganisms, preventing them from reproducing and causing infections. Integrated into the water distribution network, UV systems provide continuous disinfection without altering the water's taste or introducing additional chemicals.

- **The Radiant Solution:** UV purification is a testament to the power of light in neutralizing microbial threats. Specifically, UV-C light, with its wavelength between 200 and 280 nanometers, becomes the weapon of choice. As rainwater flows through UV systems integrated into the water distribution network, a silent yet powerful process unfolds.

- **DNA Damage:** UV-C light, when targeted at microorganisms, wreaks havoc at the molecular level. It damages the DNA of bacteria, viruses, and other pathogens, rendering them incapable of reproduction. This disruption in the life cycle ensures that even if microorganisms survive exposure to UV light, they cannot proliferate or cause infections. The result is a water supply that is continuously disinfected without altering its taste or introducing additional chemicals.

- **Integration into Water Distribution:** The seamless integration of UV systems into the water distribution network is a hallmark of their effectiveness. As rainwater makes its way through pipes and conduits, UV-C lights stand guard, providing a continuous and automated disinfection process. This integration ensures the microbial safety of the water at the point of consumption and minimizes the need for manual intervention, making UV purification a reliable and efficient safeguard.

Chlorination

Journeying into the annals of water treatment history, you'll encounter the enduring legacy of chlorination. A method that has withstood the test of time, chlorination involves the addition of chlorine to water for disinfection. Residual chlorine monitoring is a key focus, highlighting chlorination's time-tested efficacy.

- **The Enduring Legacy of Chlorine:** Chlorination is a time-tested approach with a legacy spanning over a century. The principle is simple yet effective. Chlorine, in various forms such as chlorine gas, sodium hypochlorite, or calcium hypochlorite, is a potent agent against a broad spectrum of microorganisms.

- **Broad-Spectrum Disinfection:** Chlorine's efficacy lies in its ability to eliminate not only bacteria but also viruses, algae, and other pathogens. It disrupts the life cycle by attacking cellular structures and enzymes, rendering them unable to function. The result is a comprehensive disinfection that safeguards against a wide range of potential threats in harvested rainwater.

- **Dosage Control:** While chlorination is a powerful disinfection method, the key lies in precise dosage. Adding too many compromises the taste and safety of the water, while adding too little fails to achieve effective disinfection. Achieving this balance requires careful monitoring and control of chlorine levels throughout the water distribution network.

- **Residual Chlorine Monitoring:** Maintaining residual chlorine levels becomes a key aspect of the chlorination process. Residual chlorine, the amount of chlorine that remains in the water after disinfection, is an indicator of ongoing microbial protection. Regular monitoring ensures that the water continues to meet safety standards without compromising taste. It's a delicate equilibrium that underscores the importance of chlorination's time-tested efficacy.

Boiling

As you embrace tradition in modern times, boiling takes center stage as a simple yet highly effective method of sterilizing water. Whether it's the straightforward mechanism of pathogen eradication through heat or the altitude considerations that shape boiling practices, the simplicity of boiling is the epitome of effectiveness.

- **Embracing Tradition in Modern Times:** In situations where advanced technologies aren't readily available, boiling is the best practice. Boiling, an age-old practice rooted in tradition, remains a simple yet highly effective method of sterilizing water. As rainwater reaches its boiling point of 100 degrees Celsius (212 degrees Fahrenheit), most pathogens are eradicated.

- **Pathogen Eradication through Heat:** The mechanism is straightforward. The application of heat through boiling disrupts the structural integrity of microorganisms. While boiling doesn't remove chemical contaminants, it provides a practical and accessible means of ensuring microbial safety. This simplicity becomes especially valuable in situations where access to sophisticated water treatment infrastructure is limited.

• **Altitude Considerations:** In regions with higher altitudes, where water boils at lower temperatures due to reduced atmospheric pressure, the recommended boiling time is extended to ensure complete pathogen eradication. Boiling for at least one minute (or three minutes at higher altitudes) becomes the golden rule, reaffirming the simplicity and effectiveness of this age-old practice.

From the radiant elegance of UV purification to the time-tested legacy of chlorination and the simplicity of boiling, each technique stands as a sentinel, guaranteeing that harvested rainwater reaches its final destination free from microbial threats. In this process of disinfection, tradition meets innovation, and simplicity intertwines with sophistication, creating a harmonious journey from the heavens to the human thirst.

Importance of Regular Testing and Maintenance

In the journey from raindrop to drinking glass, where advanced filtration and disinfection methods stand as guardians, regular testing and maintenance are paramount. Post-treatment testing and system maintenance ensure the sustained safety and quality of harvested rainwater. From the delicate balance of residual chlorine levels to the vigilant checks on microbial content, this iterative process is the lifeline that safeguards water quality over time.

Post-Treatment Testing

After the rainwater undergoes advanced filtration and disinfection, post-treatment testing illuminates the unseen threats that linger despite the formidable defenses of UV purification and chlorination. From the delicate balance of residual chlorine levels to the vigilant checks on microbial content and overall water quality, post-treatment testing ensures the continued safety and quality of harvested rainwater.

• **Residual Chlorine Levels:** When using the chlorination method, achieving the balance of residual chlorine levels is critical. Residual chlorine is your tell against microbial resurgence. Too little, and the water becomes vulnerable to contamination. On the other hand, with too much chlorine, the taste and safety of the water are compromised.

- **Microbial Content:** Microbial content testing delves into the microscopic universe, where unseen pathogens persist. Even the most advanced filtration methods leave behind traces of microorganisms. Post-treatment testing ensures that these invisible threats are exposed and neutralized, reinforcing the microbial safety of rainwater.

- **Overall Water Quality:** Beyond individual components, overall water quality testing provides a comprehensive evaluation. Parameters such as pH, turbidity, and dissolved oxygen contribute to the holistic understanding of water quality. Regular assessments guarantee that the water remains pleasant to the taste and free from undesirable characteristics.

- **Iterative Refinement:** The iterative nature of post-treatment testing is the key to sustaining water quality over time. It's not a one-time validation but a continual refinement process. As environmental conditions change, seasonal variations occur, and the water distribution network evolves, regular testing adapts to these dynamics, ensuring that the safety standards set for rainwater are consistently met.

System Maintenance

Filters, UV lamps, and the entire rainwater treatment system require vigilant attention to maintain their effectiveness. This section explores how regular maintenance ensures that the gatekeepers of purity continue to uphold their role in removing impurities and microorganisms. It also sheds light on the importance of periodic checks and replacements for UV lamps, those that safeguard against microbial threats.

- **Filters:** Filters, the gatekeepers of purity in the rainwater treatment process, require vigilant attention. Over time, they accumulate particles and contaminants, diminishing their effectiveness. Regular cleaning or replacement ensures that the filtration system continues to uphold its role in removing impurities and microorganisms.

- **UV Lamps:** In UV purification, the effectiveness of UV lamps is pivotal. These lamps emit powerful UV-C light that damages the DNA of microorganisms. Periodic checks and, if necessary, replacements guarantee that the UV purification system remains a formidable barrier against microbial threats.

- **System Vigilance:** Neglecting system maintenance means you're leaving the gates unguarded. As filters become clogged and UV lamps dim, the entire rainwater harvesting and treatment system becomes susceptible to a decline in effectiveness. A compromised system jeopardizes water quality and poses risks to human health.

- **The Proactive Approach:** System maintenance is not a reactive response to issues. It's a proactive approach to sustaining the integrity of the entire rainwater treatment infrastructure. Regular checks, scheduled replacements, and keeping an eye on the overall system health become the proactive measures that prevent potential problems before they compromise the safety of the drinking water.

Post-treatment testing and system maintenance embody a continual commitment to purity. It's a pledge to safeguard the journey from raindrops to drinking glass against unseen threats and system wear. In this commitment, tradition meets innovation, and simplicity intertwines with sophistication. It creates a harmonious balance that ensures sustained safety and quality of harvested rainwater for generations to come.

Understanding local water quality sets the stage for a targeted approach, while advanced filtration methods tailor the purification process for potability. Disinfection techniques, whether through UV purification, chlorination, or boiling, lead to safety. Regular testing and maintenance should be the cornerstone for sustaining the promise of safe drinking water.

Chapter 10: A Sustainable Future – Techniques in Modern-Day Conservation

In the face of escalating global challenges such as water scarcity and climate change, rainwater harvesting represents hope for a sustainable future. This final chapter explores the contemporary significance of rainwater harvesting and the latest innovations in sustainable water use. You'll discover how it integrates into broader conservation efforts. As you navigate the intricate landscape of modern-day conservation, this chapter aims to inspire you to see your rainwater harvesting endeavors as integral contributions to a global movement towards environmental stewardship.

Understanding Global Challenges

Water scarcity, once a concern limited to specific regions, has now evolved into a critical global issue. The surge in urbanization, coupled with population growth and inefficient water management practices, has placed an unprecedented strain on this planet's water resources. In the face of this growing crisis, rainwater harvesting offers a decentralized and sustainable approach to augmenting water supplies.

Water scarcity, once a concern limited to specific regions, has now evolved into a critical global issue.

Genetics4good, GFDL <http://www.gnu.org/copyleft/fdl.html>, via Wikimedia Commons: https://commons.wikimedia.org/wiki/File:Water_stress_2019_WRI.png

The catalyst for this global imperative is none other than climate change. This phenomenon has brought about significant alterations in precipitation patterns, an increase in the frequency of extreme weather events, and an exacerbation of water scarcity in various regions. To address these challenges, rainwater harvesting stands as a mitigation strategy and a resilient response to the unpredictable shifts in climate patterns.

The Escalating Crisis of Water Scarcity

The intertwining factors of rapid urbanization and population growth have drastically increased the water demand. In many regions, traditional water sources are unable to meet this surging demand, leading to water scarcity that extends beyond geographic boundaries. The urgency of the situation is magnified by inefficient water management practices that further deplete available water resources.

Unlike centralized water supply systems, which often struggle to cope with increasing demand, rainwater harvesting provides a decentralized solution. By capturing and utilizing rainwater at the local level, communities can reduce their reliance on overburdened water infrastructure and tap into a sustainable source that replenishes with each rainfall.

Climate Change

The impacts of climate change on water resources are profound and multifaceted. Altered precipitation patterns lead to irregular water availability, making it essential for communities to adapt their water management strategies. Extreme weather events, such as droughts and floods, further intensify the challenges of water scarcity, emphasizing the need for immediate and proactive measures.

By capturing rainwater, communities can build resilience against the uncertainties associated with shifting climate patterns. The decentralized nature of rainwater harvesting aligns seamlessly with the call for adaptive measures in the face of climate change.

Innovations in Sustainable Water Use

As the world grapples with the pressing challenge of water scarcity, innovative solutions are emerging to revolutionize the way people use and manage water. One such avenue of progress lies in sustainable construction and agricultural practices. Green building designs and advanced irrigation methods are reshaping the global approach to water conservation.

Green Building Designs

In the quest for sustainable water use, conservation efforts are extending beyond the conventional. Modern architects are embracing a paradigm shift by seamlessly integrating nature into building designs. Green building designs go beyond aesthetics by transforming structures into sustainable ecosystems. Rooftop gardens, permeable surfaces, and self-draining structures are all elements of a holistic approach to water conservation.

Rooftop gardens, for instance, serve a dual purpose. They enhance rainwater collection by providing a natural surface for water accumulation and contribute to urban biodiversity. These green oases create habitats for plants and insects, fostering a healthier urban ecosystem. Additionally, permeable surfaces allow rainwater to infiltrate the ground, replenishing aquifers and reducing surface runoff that can lead to flooding.

The integration of self-draining structures is another innovation in green building designs. These structures are designed to efficiently manage rainwater, directing it away from buildings and into collection systems. By doing so, they mitigate the urban heat island effect, contributing to a cooler and more sustainable urban environment.

Advanced Irrigation Methods

Agriculture, a sector that consumes a significant portion of the world's water supply, is undergoing a transformative revolution in irrigation methods. Precision agriculture, fueled by technology, is at the forefront of this change. The key objective is to deliver water precisely where and when it is needed, optimizing usage and minimizing waste.

The integration of rainwater into advanced irrigation systems further enhances their efficiency. By capturing and utilizing rainwater, farmers are reducing their reliance on traditional water sources, mitigating the impact on local ecosystems. This approach fosters a more sustainable model of food production.

Precision agriculture utilizes sensors, data analytics, and automated systems to monitor and manage crop conditions. This data-driven approach allows farmers to make informed irrigation decisions, optimizing water use for maximum crop yield. By embracing precision agriculture and incorporating rainwater into these systems, humanity is moving towards a more sustainable and water-efficient future for global agriculture.

Synergistic Environmental Benefits Through Integration

The true potential of rainwater harvesting lies in its ability to synergize with other conservation practices. This section explores the transformative impact of integrating rainwater harvesting with techniques such as greywater recycling, permaculture designs, and sustainable landscaping. It's time for you to understand how creating a holistic ecosystem extends beyond immediate water supply concerns.

Greywater Recycling

Rainwater harvesting, when seamlessly integrated with greywater recycling, forms a powerful duo in sustainable water management. Greywater, derived from daily activities such as laundry and bathing, complements rainwater by providing an additional source for non-potable uses. By combining these two sources, communities can significantly reduce their dependence on traditional water supplies, easing the burden on strained water resources.

Greywater recycling systems capture, treat, and reuse water that would otherwise go to waste. When this recycled water is synchronized with rainwater harvesting, it creates a cyclical and efficient water management system. This synergy promotes a more sustainable lifestyle, highlighting the interconnectedness of various water sources.

Permaculture Designs

Integrating rainwater harvesting with permaculture designs takes the concept of sustainability to another level. Permaculture principles guide the creation of self-sustaining environments that mimic natural ecosystems. By harmonizing rainwater harvesting with permaculture, you create landscapes that foster biodiversity, enrich soil health, and promote regenerative agriculture.

Permaculture emphasizes working with nature instead of against it. Through careful design, rainwater is directed to nourish plants, support food forests, and create microclimates that enhance overall ecosystem resilience. This approach contributes to the creation of vibrant and sustainable living environments.

Ecosystem Restoration

The impact of rainwater harvesting extends far beyond meeting immediate water needs. It becomes a catalyst for ecosystem restoration, playing a crucial role in preserving natural habitats. By replenishing groundwater levels and supporting the health of rivers and lakes, rainwater harvesting contributes to the overall well-being of ecosystems.

When integrated with other conservation practices, rainwater harvesting becomes a force for positive change. It nurtures the resilience of entire ecosystems, ensuring the health of flora and fauna that depend on sustainable water sources. This interconnected approach is a reminder that your efforts are not just about securing water for today but about creating a legacy of environmental stewardship for generations to come.

A Global Movement Toward Sustainability

In the quest for sustainability, many countries are recognizing the profound impact of their actions. Embracing rainwater harvesting as a broader conservation narrative transforms people from passive observers to active participants in a global movement toward sustainability.

Individual Actions, Global Impact

Rainwater harvesting, when embraced on an individual level, goes beyond personal water security. It becomes a cornerstone of a global movement towards sustainability. The cumulative effect of countless individuals adopting rainwater harvesting practices has a profound impact. It creates a network of interconnected efforts that transcend geographic boundaries. It influences the health of entire ecosystems and contributes to the larger narrative of environmental responsibility.

Individuals who embrace rainwater harvesting recognize that their actions are part of a broader ecosystem. By capturing and utilizing rainwater, you contribute to the conservation of traditional water sources, alleviating the strain on local water supplies. It secures water for personal use and safeguards the delicate balance of ecosystems that depend on sustainable water sources.

Community Engagement and Advocacy

The power of rainwater harvesting extends beyond its immediate utility. It catalyzes community engagement and advocacy. Individuals who have experienced the benefits of rainwater harvesting often become passionate advocates for sustainable water practices. Sharing success stories, promoting awareness, and collaborating on larger conservation initiatives create a ripple effect that amplifies the impact of rainwater harvesting.

Communities that come together to embrace rainwater harvesting initiate a positive feedback loop. As awareness spreads, more individuals are inspired to adopt these practices, creating a groundswell of support for sustainable water management. This community-driven approach strengthens local resilience and contributes to a broader cultural shift toward sustainability.

Educational Initiatives: Shaping Future Stewards

The journey toward a sustainable future involves education and empowerment. Rainwater harvesting provides an excellent opportunity to integrate sustainability into educational curricula and community outreach programs. Incorporating this practice into the learning experience will cultivate a new generation of environmental stewards.

Educational initiatives centered on rainwater harvesting go beyond theory. They provide practical knowledge that empowers individuals to make a tangible difference. As students and community members learn about the environmental impact of their choices, they become active contributors to a sustainable future. These future leaders will carry the torch forward, ensuring that the ethos of sustainability becomes an integral part of the collective consciousness.

The global movement towards sustainability is not an abstract concept but a collective effort built on individual actions. Rainwater harvesting, when embraced by individuals and communities, becomes a powerful force in this movement. From securing personal water needs to influencing the health of ecosystems and advocating for broader environmental conservation, the ripple effect of rainwater harvesting is shaping a more sustainable world. As you educate, engage, and advocate, you cultivate a legacy of environmental stewardship for generations to come.

Conclusion: A Call to Action

As you conclude this exploration into rainwater harvesting and modern-day conservation, it's essential to recognize the transformative potential within your grasp. Rainwater harvesting is more than just a technique. It's a philosophy that recognizes the interconnectedness of human actions with the health of the planet.

7 Ways Harvesting Rainwater is Beneficial for the Future

From ensuring water security to fostering ecosystem resilience, rainwater harvesting is a versatile and essential practice. Here are some ways in which integrating sustainability and conservation techniques amplifies the benefits of rainwater harvesting, creating a harmonious approach to resource management.

1. Water Security and Independence

The primary and most immediate benefit of rainwater harvesting is the assurance of water security. As populations grow and traditional water sources get drained, capturing rainwater provides a decentralized and reliable water supply. Rooftop harvesting systems, for example, allow individuals and communities to collect rainwater for various uses, from domestic needs to agricultural irrigation.

Water independence is particularly crucial in regions with unreliable infrastructure or vulnerable to drought. By harvesting rainwater, you can mitigate the impact of water shortages, ensuring a continuous and reliable water source, even in arid climates.

2. Mitigating the Urban Heat Island Effect

Urban areas often experience higher temperatures than their rural counterparts, creating what is known as the urban heat island effect. Rainwater harvesting, especially when integrated into green building designs, contributes to mitigating this heat island effect.

Green roofs and permeable surfaces, commonly associated with rainwater harvesting practices, provide shade, reduce surface temperatures, and improve overall urban microclimates. By lessening the heat island effect, rainwater harvesting contributes to creating more comfortable and sustainable urban environments.

3. Biodiversity Enhancement through Sustainable Landscaping

Rainwater harvesting extends beyond collecting water. It involves a holistic approach to landscaping that enhances biodiversity. Sustainable landscaping practices that integrate rainwater harvesting create environments that support a variety of plants and animal life.

Utilizing rainwater for landscaping reduces the reliance on traditional irrigation methods, conserving water and fostering a healthier ecosystem. Native plants, adapted to local climates, thrive with rainwater, attracting diverse wildlife and contributing to the preservation of local biodiversity.

4. Soil Health and Regenerative Agriculture

Rainwater harvesting plays a crucial role in promoting soil health and regenerative agriculture. By capturing rainwater and directing it to agricultural fields, farmers reduce their dependence on unsustainable water sources and adopt more environmentally friendly irrigation practices.

The replenishment of soil moisture through rainwater harvesting contributes to improved soil structure and fertility. This, in turn, supports sustainable farming practices, reduces soil erosion, and enhances the overall resilience of agricultural ecosystems.

5. Mitigating Flood Risks and Stormwater Management

In urban areas, heavy rainfall often leads to flooding and strain on stormwater management systems. Rainwater harvesting acts as a natural solution to mitigate these risks by reducing surface runoff.

When rainwater is harvested and used on-site, less water enters stormwater drains, lowering the risk of flooding. Additionally, the process of collecting rainwater helps filter out impurities, reducing the burden on stormwater management infrastructure and improving water quality.

6. Energy and Cost Savings

Rainwater harvesting leads to energy and cost savings. Traditional water supply systems, which involve pumping and treating water for distribution, consume significant amounts of energy. Using locally harvested rainwater decreases the demand for centralized water supply, resulting in reduced energy consumption and lower utility costs.

7. Drought Preparedness and Climate Resilience

As climate change leads to more frequent and severe droughts, rainwater harvesting becomes a vital tool for drought preparedness and climate resilience. By capturing rainwater during times of plenty, communities can build reservoirs for use during drier periods.

This proactive approach to water management contributes to climate resilience, ensuring a more reliable water supply in the face of changing weather patterns. Rainwater harvesting acts as a buffer against the uncertainties associated with climate change.

As individuals, communities, and societies, the commitment to rainwater harvesting is a commitment to a sustainable future. It is an acknowledgment that every drop saved today is a gift to the generations of tomorrow. The journey toward sustainability is a collective endeavor where each raindrop harvested becomes a symbol of hope, resilience, and the promise of a thriving planet for all.

Here's another book by Dion Rosser that you might like

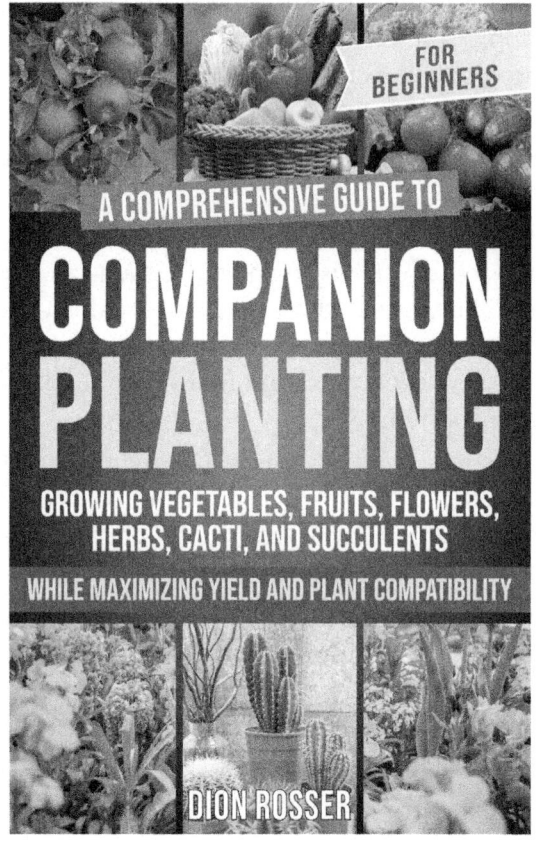

References

Cement, J. K. (2023, August 11). Rainwater Harvesting Methods, Techniques, and Tips. JK Cement. https://www.jkcement.com/blog/construction-planning/rain-water-harvesting-techniques/

Components of a Rainwater Harvesting System. (n.d.). Rainwaterharvesting.org. http://www.rainwaterharvesting.org/Urban/Components.htm

Harvesting Rainwater. (n.d.). Rainwater harvesting for drylands and beyond by Brad Lancaster. Rainwater Harvesting for Drylands and Beyond by Brad Lancaster. https://www.harvestingrainwater.com/

Housing News Desk. (2023, June 12). Rainwater Harvesting: Importance, Techniques, Pros, and Cons. Housing News. https://housing.com/news/different-rain-water-harvesting-methods/

Maxwell-Gaines, C. (2004, April 3). Rainwater Harvesting 101. Innovative Water Solutions LLC. https://www.watercache.com/education/rainwater-harvesting-101

Ogale, S. (2023). Rainwater Harvesting System. In Encyclopedia Britannica.

Rainwater Harvesting System: Steps, advantages & types. (n.d.). Ultratechcement.com. https://www.ultratechcement.com/for-homebuilders/home-building-explained-single/descriptive-articles/the-steps-to-an-efficient-rainwater-harvesting-system

Rainwater Harvesting. (2016, January 6). BYJUS; BYJU'S. https://byjus.com/biology/rainwater-harvesting/

Ruchi Singhal case study. (n.d.). Rainwater Harvesting. Cseindia.org. https://www.cseindia.org/rainwater-harvesting-1272

Sarkar, S. K., & Tigala, S. (2022, October 27). Harvest Rainwater for Water Security. BusinessLine. https://www.thehindubusinessline.com/opinion/harvest-rainwater-for-water-security/article66060897.ece

Vartan, S. (2020, December 4). A Beginner's Guide to Rainwater Harvesting. Treehugger. https://www.treehugger.com/beginners-guide-to-rainwater-harvesting-5089884

Water conservation : Rainwater Harvesting. (n.d.). Mygov.In. https://blog.mygov.in/water-conservation-rainwater-harvesting/

Printed in Great Britain
by Amazon

42386135R00076